Research on Intelligent Manufacturing

Research on Intelligent Manufacturing (RIM) publishes the latest developments and applications of research in intelligent manufacturing—rapidly, informally and in high quality. It combines theory and practice to analyse related cases in fields including but not limited to:

Intelligent design theory and technologies
Intelligent manufacturing equipment and technologies
Intelligent sensing and control technologies
Intelligent manufacturing systems and services

This book series aims to address hot technological spots and solve challenging problems in the field of intelligent manufacturing. It brings together scientists and engineers working in all related branches from both East and West, under the support of national strategies like Industry 4.0 and Made in China 2025. With its wide coverage in all related branches, such as Industrial Internet of Things (IoT), Cloud Computing, 3D Printing and Virtual Reality Technology, we hope this book series can provide the researchers with a scientific platform to exchange and share the latest findings, ideas, and advances, and to chart the frontiers of intelligent manufacturing.

The series' scope includes monographs, professional books and graduate textbooks, edited volumes, and reference works intended to support education in related areas at the graduate and post-graduate levels.

If you are interested in publishing with the series, please contact Dr. Mengchu Huang, Senior Editor, Applied Sciences

Email: mengchu.huang@springer.com Tel: +86-21-2422 5094.

More information about this series at http://www.springer.com/series/15516

Dinghua Zhang · Ming Luo · Baohai Wu ·
Ying Zhang

Intelligent Machining of Complex Aviation Components

华中科技大学出版社
Huazhong University of Science and Technology Press

Springer

Dinghua Zhang
Northwestern Polytechnical University
Xi'an, Shaanxi, China

Ming Luo
Northwestern Polytechnical University
Xi'an, Shaanxi, China

Baohai Wu
Northwestern Polytechnical University
Xi'an, Shaanxi, China

Ying Zhang
Northwestern Polytechnical University
Xi'an, Shaanxi, China

ISSN 2523-3386 ISSN 2523-3394 (electronic)
Research on Intelligent Manufacturing
ISBN 978-981-16-1585-6 ISBN 978-981-16-1586-3 (eBook)
https://doi.org/10.1007/978-981-16-1586-3

Jointly published with Huazhong University of Science and Technology Press
The print edition is not for sale in China (Mainland). Customers from China (Mainland) please order the print book from: Huazhong University of Science and Technology Press.

This Springer imprint is published by the registered company Springer Nature Singapore Pte Ltd.
The registered company address is: 152 Beach Road, #21-01/04 Gateway East, Singapore 189721, Singapore

Preface

Theoretically, as long as the correct part model is used in NC machining to generate the "correct" program, the qualified workpiece can be produced. However, in practical production, especially for the machining of complex thin-walled parts in aviation, the NC machining process is not always in an ideal state. Material removal will cause a variety of complicated physical phenomena, such as machining geometric errors, thermal deformation, elasticity deformation, system vibration, etc. The existence of these problems makes the "correct" program generated according to the correct theoretical model not necessarily able to manufacture qualified and high-quality parts. At the same time, the processing capacity of the equipment is not fully utilized, and the service life of machine tool components and cutters will also be affected. The reason for the above-mentioned problems is that the traditional machining process often only considers CNC machine tools or the machining process itself, lacking a comprehensive understanding of the interaction mechanism between the machine tool and the machining process, and it is difficult to accurately model the machining process system in advance. This kind of interaction usually comes with unpredictable effects, which greatly increases the difficulty of processing control and makes it harder to achieve precise control of the machining process. The structure of blisks, casings and other parts on complex equipment like aeroengines is becoming more and more complex, and their extremely harsh service environment has higher and higher requirements for machining process and quality. On the basis of more than 30 years of practical experience and research in the manufacturing of aviation complex thin-walled parts, the author has conducted systematic research on intelligent machining technologies in the past 8 years and proposes a relative general framework and implementation methods. Some primary research results include:

(1) "No trial cutting" detection processing method is proposed. Through the combination of active excitation and online monitoring, trial cutting is integrated into the part machining process to ensure that the workpiece material, structure and process are exactly the same in trial cutting and actual processing, and solve the problem of model inaccuracy caused by different modeling conditions and machining process in existing process models.

(2) Autonomous learning and model evolution based on detection processing is proposed. Utilizing online detection to obtain real-time operating conditions and system response information, the accumulation of process knowledge and model evolution is achieved according to the establishment of a mapping relationship between the associative memory knowledge template representing working conditions, interface coupling behaviors and workpiece quality. Aiming at the strong time-varying characteristics of the workpiece state and tool wear during the machining process, the dynamic modeling of the workpiece and tool state in the process step is realized through the time-space subdivision polymorphic evolution modeling method. Using the data storage template between the steps, iterative learning and evolution of the comprehensive machining error compensation model are achieved based on the on-site measurement and offline detection, which also solves the problem that the existing process models and modeling methods are difficult to implement dynamic modeling, independent learning and adaptive evolution.

(3) Residual stress-induced deformation of perception and prediction mathematical model is established. Adopting the statically indeterminate theory, a solution method of the residual stress-induced deformation perception and prediction model based on clamping force monitoring is proposed, which provides a new idea for the on-site deformation prediction of complex thin-walled aviation parts.

A series of related models and methods have been applied to the manufacturing of large-scale aeroengine fan blades, blisks, casings and structural components, and have achieved excellent effects.

When this book is completed, the author sincerely thanks all academic predecessors, teachers and colleagues for their support and help. This book is based on the research results of doctoral students under the guidance of the author, including Xu Zhou, Feiyan Han, Yilong Liu, Yongfeng Hou, Yaohua Hou, Ce Han, Jiawei Mei, Junjin Ma, Junteng Wang, Dongsheng Liu, Zhongxi Zhang, Qi Yao, etc. I also express gratitude to them.

This study was co-supported by the National Key Basic Research and Development Program (2013CB035802), the National Natural Science Foundation of China (51305354, 51475382, 5157553, 51675438, 91860137, and 52022082).

Special statement: This book does not have a unified symbol table, and the symbol definitions of each chapter are self-contained.

Xi'an, China

Dinghua Zhang
Ming Luo
Baohai Wu
Ying Zhang

Contents

Chapter 1
Introduction

1.1 Numerical Control Machining Technology

1.1.1 Development of Numerical Control Technology

Numerical Control Technology is a tech for automatically controlling a particular machining process by means of digital control. In 1948, United States Air Force adopted the template equipment for the development of helicopter propeller blade profile inspection. On account of the complex shape of the sample and high precision requirement, the idea of using digital pulse to control the machining tools was put forward. In 1952, the Massachusetts Institute of Technology (MIT) and Parsons technology co-researched the first three-coordinate CNC milling machine using electronic tube components trial production has been successfully developed. In 1959, numerical control devices using transistor components and printed circuit boards appeared, at the same time, automatic tool changing device came to the true, thus the machining center was born. From the late 1960s to the mid-1970s, the computerized numerical control system controlled by small-size computers and the microcomputer numerical control system using microprocessors and semiconductor memories emerged successively. Furthermore, from the 1980s, the appearance of the automatic programming technology of human–machine dialogue and PC + CNC system greatly promoted the development and application of the NC machining technology. Nowadays NC technology, also known as CNC technology, is the tech using computer software to realize the data storage, processing, computing, logical judgment and other complex functions [1, 2].

Under ideal conditions, the NC machine tool moves according to the planned processing path which is compiled on the basis of the theoretical model of the part, and qualified machined parts can be obtained. However, some product quality problems often appear during the processing such as using the same CNC machining program

D. Zhang et al., *Intelligent Machining of Complex Aviation Components*, Research on Intelligent Manufacturing, https://doi.org/10.1007/978-981-16-1586-3_1

and equipment but getting different machining quality or unstable product quality, geometrically qualified parts dissatisfying the service performance requirements. The reason for the appeal problem is that the machining process of the machine tool is not always in an ideal state. With the removal of materials, a variety of complex physical phenomena often occur, such as machine tool motion errors, thermal deformation, elastic deformation, and system vibration. The existence of these phenomena leads to the difference between the actual machining process and the ideal processing state of the part and affects the final product quality. In the past research and actual production, we usually only pay attention to the CNC machine tool or the machining process itself, lacking a comprehensive understanding of the interaction mechanism between the machine tool and machining process, and this interaction often produces unpredictable effects, which greatly increases the difficulty of machining process control [3]. In fact, the above influence is not obvious for the parts with relatively simple structure, low precision requirements and large wall thickness. And for parts with complex structure and small wall thickness, the processing course and product quality are mainly controlled by experienced workers at the present stage.

However, the structure of components on complex equipment such as aero-engine is becoming more and more complicated, and the processing quality of such products has a significant impact on their service performance.For instance, the machining accuracy of aero-engine compressor blades has a direct impact on its aerodynamic performance, and the stress state, as well as the surface microstructure of the workpiece, play a decisive effect on its fatigue life. This puts forward higher requirements for the quality stability and consistency of the processing of complex parts. Therefore, the traditional quality assurance methods which only focus on the inspection evaluation of machining results and only consider the geometrical machining accuracy cannot meet the requirements of the new generation of high-end equipment [4].

In the machining process, the tool cutting materials generates force, heat and other processing loads [5], which lead to the vibration, deformation and other responses on the machining equipment, these responses, in turn, affect the machining process and surface quality, thus forming the complicated machining process system. The structure of this process system is highly complex, non-linear and hard to model accurately [6]. At the same time, there are a large number of uncertainties and random factors in the processing, which makes it more difficult to predict the response and quality of the machining course [7]. For example, in the cutting of difficult-to- materials such as aero-engine superalloy, under the action of severe mechanical and thermal coupling, the tool wears [8] rapidly and may need to be replaced after machining for dozens of minutes [9], as shown in Fig. 1.1. Furthermore, because of the randomness of tool wear itself, it is hard to predict the accurate tool wear value, but replace the cutter in advance according to experience, which means the tool cannot be effectively used, resulting in the high cost of cutting tools, and the non-uniformity and performance fluctuation of materials which further lead to the advanced tool destruction and machined surface failure.

In summary, the state of traditional processing technology is no longer compatible with the manufacturing quality requirements of high-end equipment parts, and the

Fig. 1.1 Tool wear in milling of aviation nickel-based superalloy [9]

complexity of the processing technology system contradicts the high consistency and quality requirements of the processing for such products. To solve the above problems, it is the must to change the traditional concept and combine the machine tools and the machining process with the interaction of modeling and simulation to realize the optimization of the machining process, the improvement of the system design and the reduction of machining defects [4]. At the same time, with the help of advanced sensor technology and other related CNC equipment, the machining condition can be perceived and predicted in time and the machining parameters and status can be evaluated and adjusted to improve the shape accuracy and surface quality economically and effectively [10]. Therefore, it is urgent to break through the bottleneck of existing processing technology, develop a new generation of processing technology, and meet the needs of major equipment development.

In recent years, with the development of sensing and monitoring technology, computing technology, data processing and artificial intelligence, it has become possible to adopt new technologies to solve the problems of accurate modeling of complex process systems and accurate prediction and control of processing system responses. The combination of these emerging technologies and processing also promotes the development of intelligent processing, a new generation of processing technology.

1.1.2 Development Stage of CNC Machining Model

With the development of NC technology, NC processing models are also constantly evolving, from the initial processing of only two-dimensional graphics to consideration of geometric and physical constraints, as well as the current stage of intelligent processing, as shown in Fig. 1.2. The basic characteristics of each stage are as follows.

(1) Geometric model stage: This stage is mainly to solve the geometric and path control problems in the machining process according to the geometric characteristics of the part, generate the CNC machining toolpath of the product, and realize the automatic high-precision machining. The main content of the

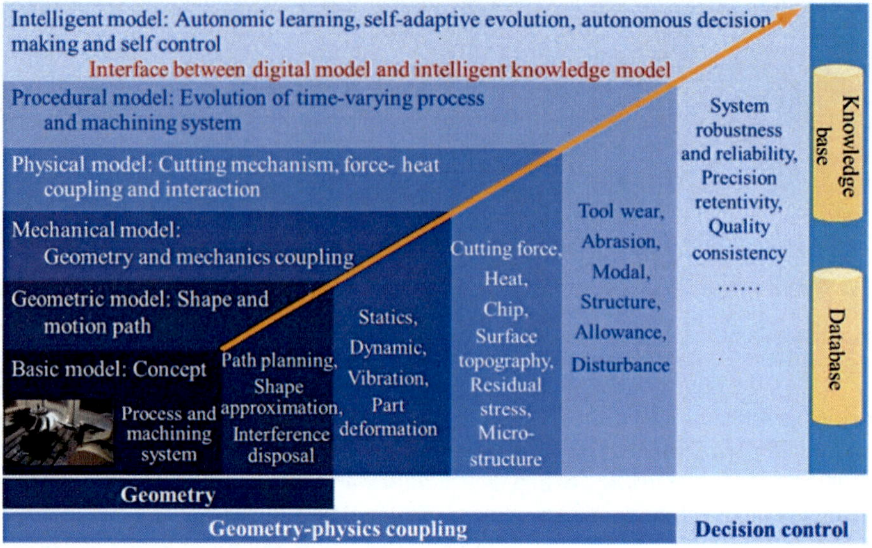

Fig. 1.2 Several development stages of the CNC machining model

research at this stage includes the planning of the machining path, the accurate approximation of the machining path and the geometry of the workpiece, and the treatment of the interference between the tool and the workpiece. This stage occupies a long time in the development of CNC machining models. With the changes brought by the third industrial revolution, the basic problems of automated manufacturing and mass production of complex parts have been solved.

(2) Mechanical model stage: This stage is mainly to solve the problems of statics and dynamics in the process of complex parts. The main research is to control the deformation of the tool, the vibration suppression of the machine tool and the workpiece during processing, and further improve the machining accuracy and efficiency of the workpiece. This stage of development probably started in the early 1980s and continues to the present, and is an important research field in thin-walled parts processing. The development at this stage made people realize the limitation of only considering geometric problems in CNC machining, and developed related control methods, which greatly improved the machining efficiency and accuracy.

(3) Physical model stage: This is a long-lasting stage in the research of machining technology. The research of this stage aims to start from the cutting mechanism and integrate the research results into the processing of parts. Since the beginning of the twentieth century, there have been a lot of researches on the cutting mechanism of metal materials. At the stage of rapid development of NC technology, the application of these basic theories in engineering has promoted the further improvement of workpiece processing quality. At present, this stage

is still in continuous development and focuses on the formation mechanism of the surface microstructure and residual stress of the workpiece during the machining process and the influence of the thermo-mechanical coupling on the surface integrity and the service performance of the parts.

(4) Procedural model stage: On the basis of studying the single-point problem at a fixed time in the machining process, the change of the system state during the processing and its influence on the processing quality has been one of the research focuses in the machining of complex components in the past two decades. The research contents mainly include the influence of material removal of thin-walled parts on their dynamic characteristics, the influence of different poses of large-scale CNC machine tools on the machining process, the influence of performance changes on the machining quality during long-term use of the machine tool, and the influence of time-varying factors such as tool wear. At this stage, the machining process is regarded as a complex dynamic process, which plays an important role in meeting the high-quality manufacturing requirements of high-end equipment complex components.

(5) Intelligent model stage: With the increase in the structural complexity of high-end equipment parts and higher requirements for manufacturing quality, the realization of accurate control of the machining process, accurate guarantee of the surface quality, and the correct processing of each product have become important for the development of processing technology. The intelligent model stage aims to realize the self-learning, self-evolution, and self-decision for the machining process through the application of related technologies, thereby improving the robustness and reliability of the processing technology system, and comprehensively improving the consistency of product processing quality.

At present, in the context of increasingly intensified global technological competition and re-emergence of technological barriers, accelerating the transformation and upgrading of technological development in the manufacturing sector, controlling the core technology of high-value-added product manufacturing, and developing data-driven intelligent processing technology has become the technology in the manufacturing field of all countries in the world.

1.2 Intelligent Processing Technology

1.2.1 Intelligent Processing Technology

Intelligent machining technology belongs to the category of manufacturing process intelligence, with the goal of realizing the intelligence of CNC machining process, including key technologies such as CNC machining process system and process modeling, optimization control and intelligent machining system integration, as well as intelligent machine tools and intelligent fixtures and intelligent network communication [3].

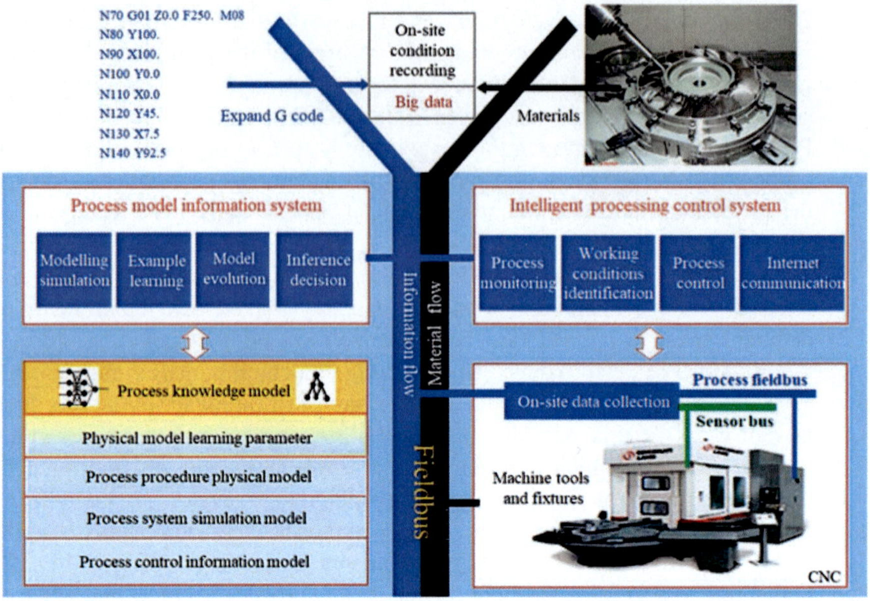

Fig. 1.3 The overall structure of the intelligent processing system

As shown in Fig. 1.3, the intelligent processing system is composed of a process model information system, an intelligent processing control system, a processing knowledge base, and equipment. It is the "middle layer" of the intelligent manufacturing system's vertical integration from the workshop to the bottom equipment, and also a system that integrates the underlying process information model and physical processing equipment. Through modern sensing, network communication and other related technologies, the system can realize the dynamic integration of intelligent processing control system and intelligent processing equipment (dynamic monitoring and control of processing system, procedure and equipment), and knowledge integration based on process model information system (data storage and processing, knowledge mining, integration and management).

1.2.2 Ways to Realize Intelligent Processing

In intelligent processing systems, advanced digital testing and processing equipment and virtual simulation methods are generally used to achieve modeling, simulation, prediction, and optimization of the process, as well as online monitoring and control of the real process, that is, to achieve dynamic integration. In addition, the integration of existing process knowledge and reasoning decision-making mechanism enable the processing system to automatically optimize machining parameters according to

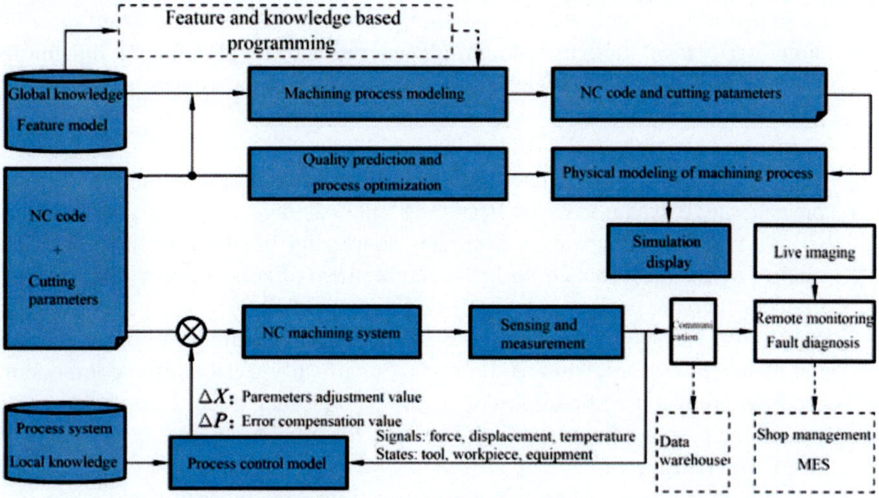

Fig. 1.4 The overall technical route of the intelligent processing

real-time machining conditions, adjust its own state, and obtain optimal processing performance and quality efficiency, that is, to achieve knowledge integration.

A typical intelligent processing technology route is shown in Fig. 1.4. According to different parts of the machining process planning and various parameters such as feed rate, etc., which have an effect on product quality and processing efficiency, by the means of the simulation model based on machining process, cutting parameters can be predicted and selected, furthermore, generating optimal machining process control instructions. During the machining process, various sensors, remote monitoring and fault diagnosis technologies are used to detect the vibration, cutting temperature, tool wear, machining deformation and the running state and health of the equipment. On this basis, according to the pre-established system control model, the machining parameters can be adjusted, and the errors generated in the machining process can be compensated in real time.

It can be seen from the connotation and implementation approach of the above intelligent processing technology that intelligent processing technology is a cross-technology involving CNC machining, cutting, sensing, control and other fields.

1.2.3 Basic Knowledge of Intelligent Processing Technology

1. **Numerical Control Machining**
   CNC machining technology mainly involves the use of programming software to automatically generate machining toolpaths according to the structural characteristics and process requirements of the parts, and to control CNC machine tools to make them move in accordance with the planned path. The

main research contents include: planning of machining toolpath, avoiding collision interference, toolpath smoothing, generation of highly efficient machining path, feature-driven intelligent programming and digital control of machining process.

2. **Cutting**
   Cutting process is a method to use the relative motion of cutting tool and workpiece, remove excess material from the blank (castings, forgings, profiles, etc.), so as to obtain dimensional accuracy, shape and position accuracy, surface quality in full accordance with the requirements of the pattern. Most of the blanks which are processed by casting, forging and welding, are generally difficult to use and assemble, they need to be further machined, in order to meet the technical requirements of the parts. Specifically, aero-engine components are high precision and complex structure parts, which need to be machined to ensure the processing accuracy and surface machining quality.
3. **Big Data Mining**
   Data mining is a series of procedures whose ultimate purpose is to extract desired or unexpected information from data. A large amount of data will be generated during the machining process, including the force, heat, vibration, deformation and other data generated by the tool's continuous cutting materials, the machine tools' operation position, poses, vibration, as well as the workpieces' geometric size, surface status, tool wear and other data after machining. When these data are collected, the complete data reflecting the machining process will be obtained. These data are of large volume, variety and strong correlation. Therefore, clustering analysis and other data mining methods can be used to analyze the machining process and establish the correlation between the machining process data and the surface machining quality, efficiency and even the service performance of the workpiece.
4. **Machine Learning**
   Machine learning is a multidisciplinary interdisciplinary subject, involving probability theory, statistics, approximation theory, convex analysis, algorithm complexity theory and other disciplines. Machine learning in the machining process is mainly to make the processing technology system gradually improve its performance with the accumulation of experience knowledge and data, and continuously enhance the processing quality and efficiency of products. A lot of data mining methods will be used in machine learning.
5. **Intelligent Control**
   In the traditional NC machining, the main function of the NC machine tool is to control the machine to execute the corresponding movement according to the given processing code. The abnormal conditions, such as machining vibration, deformation and tool failure, are generally not monitored and controlled. However, with the improvement of the processing quality requirements of high-end equipment, the monitoring and control of abnormal processing conditions

is of vital importance to ensure the processing quality of products. Typical intelligent control of machining process includes monitoring and active control of machining vibration and deformation, monitoring and correction of residual stress induced deformation, monitoring and online regulation of the abnormal state, etc.

1.3 The Content of This Book

This book is a systematic summary of the author's regular understanding and innovative achievements in the fields of machining process modeling, process model, machining process optimization and control. The rest of the book is arranged as follows:

Chapter 2 introduces the polymorphic evolution process model of the time-varying machining process, including the description of the machining process system, polymorphic evolution model of the machining process, geometric and dynamic evolution model of workpiece and tool wear evolution model, etc.

Chapter 3 introduces the monitoring and data processing methods of the machining process, including the detecting methods in the machining process, cutting depth and width and tool wear state, and identifying methods of cutting force system based on field monitoring data, etc.

Chapter 4 introduces the learning optimization method of the process model, including the basic principle of process learning optimization, the spatio-temporal mapping method of process data, the compensation iterative learning method of machining error, the iterative optimization method of deep-hole drilling depth and the cyclic iterative optimization method of machining parameters, etc.

Chapter 5 introduces the dynamic response prediction and control methods of the machining process. Taking typical engine casing and blade as examples, the dynamic response prediction of machining process, the dynamic response control method based on cutting parameter optimization and local clamping enhancement, and the control method of damping auxiliary support are described.

Chapter 6 introduces the clamping perception method of residual stress-induced deformation in thin-walled parts, including the generation of machining-induced residual stress in thin-walled parts, the perception and prediction principle and method of residual stress induced deformation, etc.

References

1. ZHOU J, ZHOU Y H. Numerical control machining technology [M]. Beijing: National Defense Industry Press, 2002.
2. LIU X W, ZHANG D H, WANG Z Q, et al. Numerical control machining theory and programming technology [M]. Beijing: China Machine Press, 2001.

3. ZHANG D H, LUO M, WU B H, et al. Development and application of intelligent machinigtechnology [J]. Aeronautical Manufacturing Technology, 2010, 21: 40–43.
4. M'SAOUBI R, AXINTE D, SOO S L, et al. High performance cutting of advanced aerospace alloys and composite materials [J]. CIRP Annals–Manufacturing Technology, 2015, 64(2): 557–580.
5. CUI D, ZHANG D H, WU B H, et al. An investigation of tool temperature in end milling considering the flank wear effect [J]. International Journal of Mechanical Sciences, 2017, 131–132: 613–624.
6. HOU Y F, ZHANG D H, WU B H, et al. Milling force modeling of worn tool and tool flank wear recognition in end milling [J]. IEEE-ASME Transactions on Mechatronics, 2015, 20(3): 1024–1035.
7. ZHOU X, ZHANG D H, LUO M, et al. Toolpath dependent chatter suppression in multi-axis milling of hollow fan blades with ball-end cutter [J]. The International Journal of Advanced Manufacturing Technology, 2014, 72(5): 643–651.
8. WANG J R, LUO M, XU K, et al. Generation of tool-life-prolonging and chatter-free efficient toolpath for five-axis milling of freeform surfaces [J]. Journal of Manufacturing Science and Engineering, 2019, 141(3): 1–15.
9. LUO M, LUO H, ZHANG D H, et al. Improving tool life in multi-axis milling of Ni-based superalloy with ball-end cutter based on the active cutting edge shift strategy [J]. Journal of Materials Processing Technology, 2018, 252: 105–115.
10. LUO M, LUO H, AXINTE D, et al. A wireless instrumented milling cutter system with embedded PVDF sensors [J]. Mechanical Systems and Signal Processing, 2018, 110: 556–568.

Chapter 2
Polymorphic Evolution Process Model for Time-Varying Machining Process

For the description of the processing technology system, most commercial softwares, when defining the processing technology system, describes the complete machine structure, tooling and fixture, numerical control system and the relationship among them. Some other software focus more on the establishment of the numerical control system itself. The process system description for intelligent machining consists of two parts, namely, the equipment and hardware components, the interaction between them and the data, as well as the physical part and the digital part. As most aviation complex structure parts are thin-walled and easy to be deformed in the machining. In the machining process system construction studied in this book, the process system including the thin-walled workpiece is mainly considered, and the cutters-spindle subsystem and workpiece-fixture subsystem are mainly studied, as well as the interaction and state transfer between cutters and workpiece in the milling process.

2.1 Description of the Machining Process System

In Fig. 2.1, the process system can be divided into three levels from the inside out. The innermost layer is the cutter-workpiece interaction interface, where the cutter cuts the workpiece into chips to perform the most basic cutting operations. The middle layer is the subsystem layer, which mainly includes the cutter-spindle subsystem and workpiece-fixture subsystem. The outermost layer is the process system layer, which contains the process equipment and its system's response during the processing[1].

During the machining process, significant part deflection and cutting vibration occur due to the action of periodic alternating cutting forces, which lead to system instability, large machining error and lower surface quality. And this phenomenon is especially significant in the processing of thin-walled parts. In the machining

D. Zhang et al., *Intelligent Machining of Complex Aviation Components*, Research on Intelligent Manufacturing, https://doi.org/10.1007/978-981-16-1586-3_2

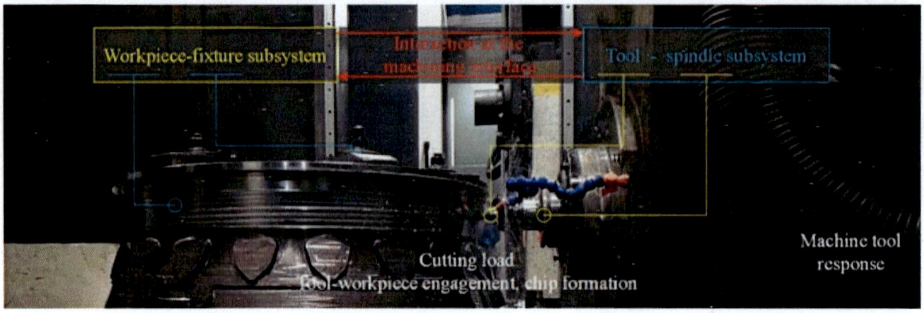

Fig. 2.1 Process system composition

process, the cutter-spindle subsystem and the workpiece-fixture subsystem interact on the cutting interface and stimulate the dynamic response of the system [2].

Taking the milling process as an example, the basic dynamic equation of the machining system is:

$$\boldsymbol{M}\ddot{\boldsymbol{X}} + \boldsymbol{C}\dot{\boldsymbol{X}} + \boldsymbol{K}\boldsymbol{X} = \boldsymbol{F}(t) \tag{2.1}$$

The main task of the milling process dynamics modeling is to establish the model according to the parameters of the milling forces in Eq. (2.1), modal mass $\boldsymbol{M}$, modal damping $\boldsymbol{C}$ and modal stiffness $\boldsymbol{K}$. Moreover, to model the milling force needs to set up the milling expression and determine the parameters of milling force according to the actual model, and to model the modal parameters is mainly determined by the experimental modal analysis method. Based on the establishment of the dynamic model of the process system, the dynamic response and stability prediction of the process system can be predicted, as shown in Fig. 2.2.

During machining process, the state of the cutter and the workpiece is constantly changing. The major change of the cutter state is mainly about the change of the wear amount in the machining process, for which the method of forming element slices along the axis is mainly adopted to analyze. Meanwhile, the change of the workpiece state is mainly manifested in the machining allowance along the machining process

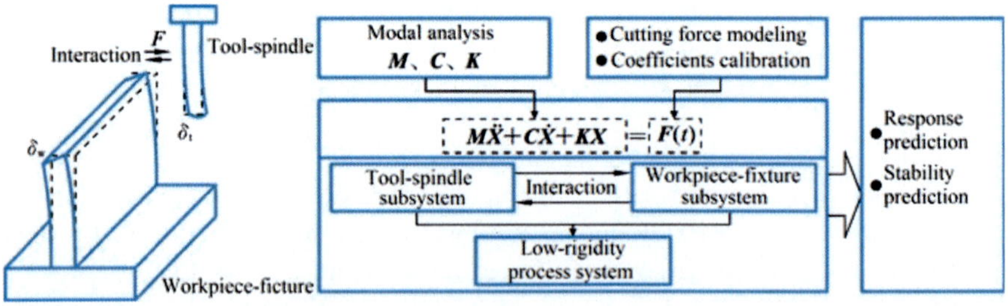

Fig. 2.2 Dynamic modeling and analysis flow of thin-wall parts milling process system

and toolpath changes, and the resulting workpiece dynamics properties change, it also reflects the effect of spatial and temporal variations of the cutting process [1].

2.1.1 The Cuter-Spindle Subsystem Dynamics Model

Among the components of the machining process system, the cutter-spindle system is relatively independent and fixed. When the vibration along the cutter axis is not considered and other influences of the machine tool body are ignored, the subsystem formed by the cutter-spindle can be simplified to the mass-spring-damping system with multiple degrees of freedom (DOF). The corresponding dynamic equation is as follows:

$$\boldsymbol{M}_T\ddot{\boldsymbol{X}} + \boldsymbol{C}_T\dot{\boldsymbol{X}} + \boldsymbol{K}_T\boldsymbol{X} = \boldsymbol{F}_T(t) \tag{2.2}$$

where $\boldsymbol{M}_T$, $\boldsymbol{C}_T$, and $\boldsymbol{K}_T$ are the mass matrix, damping matrix and stiffness matrix of the cutter-spindle subsystem respectively. The cutter-spindle subsystem can be regarded as a linear time-invariant system without considering tool wear and machine tool motion.

As shown in Fig. 2.3, when the tool direction and position change, the modal parameters of the machine tool system will also change. At this point, the full mode field corresponding to the tool tip in the machine tool system can be established to describe the tool tip mode under different positions of the machine tool:

$$\begin{cases} \boldsymbol{M}_T = \boldsymbol{M}(x, y, z, \alpha, \beta) \\ \boldsymbol{C}_T = \boldsymbol{C}(x, y, z, \alpha, \beta) \\ \boldsymbol{K}_T = \boldsymbol{K}(x, y, z, \alpha, \beta) \end{cases} \tag{2.3}$$

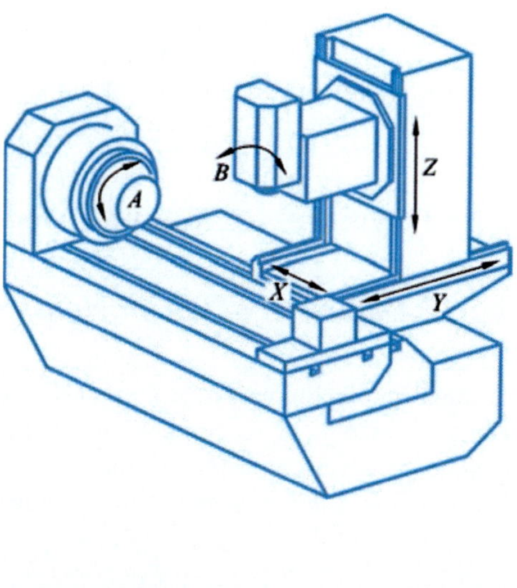

Fig. 2.3 Five-axes NC machine tool

where x, y and z are the translational coordinates of the machine tool; α, and β correspond to the two rotational coordinates of the machine tool. After the machining trajectory is determined, according to the full mode field of the machine tool, the motion coordinates of the machine tool can be expressed as a function of the arc length parameter s of the tool tip, such as $x = x(s)$. It should be noted that in the actual machining, the change of the angle direction of the rotation axis does not necessarily have a significant impact on the dynamic characteristics of the machine tool, and usually, only the motion coordinates with the most significant impact need to be recorded. The dynamic characteristics of multi-axis machine tools can be stored in C space [3, 4].

2.1.2 Workpiece-Fixture Subsystem Dynamic Model

In the low rigid process system composed of thin-walled parts, the low rigidity of the workpiece is dominant and has a significant influence on the response of the process system under the interaction. When the vibration of the workpiece is considered separately, at time t, it is reduced to a two-degree-of-freedom mass-spring-damping system, and its dynamic equation can be obtained:

$$\boldsymbol{M}_W(t)\ddot{\boldsymbol{X}} + \boldsymbol{C}_W(t)\dot{\boldsymbol{X}} + \boldsymbol{K}_W(t)\boldsymbol{X} = \boldsymbol{F}_W \tag{2.4}$$

During the machining process, the physical parameters of the workpiece, including modal mass and modal stiffness, change with the removal of the blank material and the change of the machining position. Therefore, the time-varying dynamic parameters in the above model reflect the time-varying characteristics of thin-walled parts processing. Among them, the time-varying parameters in the model can be expressed by polynomial fitting to the sampling cutting position parameter data and can be stored in the 3D volume element of the workpiece by the time-position mapping method in Sect. 4.2. In addition, because the workpiece itself is relatively thin, the workpiece and the fixture generally constitute the workpiece-fixture subsystem for analysis and control in the analysis process.

2.2 Polymorphic Evolution Model of Machining Process

2.2.1 Definition of the Machining Process

During the machining process, with the cutting of the workpiece material and the constant change of the processing position, the process system state constantly changes, and this constantly changing process becomes the machining process. For the convenience of analysis, the actual continuous machining process can be discretized into several ordered moments, each of which has its corresponding process system state, and the set of these discrete space–time states is defined as polymorph [5].

For the process of thin-walled parts, the dynamic characteristic changes of the process system correspond to the dynamic characteristic changes of the work-fixture subsystem and cutter-spindle subsystem respectively. In terms of the workpiece-fixture subsystem, the dynamic characteristics of thin-walled parts change constantly as the workpiece material is removed during the machining process, which leads to the constant change of the low rigidity workpiece system and increases the difficulty of the machining process control of thin-walled parts. At present, there are two main methods to describe the state of the workpiece, that is, the design model of the workpiece and the blank model before processing, but the description method of the state of the workpiece in the machining process is not perfect. Therefore, it is particularly important to describe the machining process of thin-walled parts and the different states of the workpiece in the machining process with the appropriate model. As for the cutter-spindle subsystem, the friction state between the cutter's back surface and the machined surface of the workpiece changes with the cutter's constant wear during the cutting process, which leads to the constant change of the components of the milling force and changes the response of the low rigidity machining process system to the interaction. The polymorphic evolution model is the discrete representation of the processing process in the time domain, including the polymorphic model and evolution model.

The polymorphic model is the state set of the machining process on the discrete-time series, and the evolution model is used to describe the transition process between adjacent states in the polymorphic model. By establishing a polymorphic evolution model, the machining process and the changes of workpiece and tool can be described. In addition, the time-domain discretization can transform the nonlinear time-varying process into a local linear time-invariant problem for modeling and solving.

2.2.2 Time Domain Dispersion of the Machining Process

In the machining process, the cutting position of the tool on the workpiece is constantly changing, and the machining process can be discretized according to the time series at different levels, such as the cutting position of the tool, the cutting state of the workpiece material, and the working procedure stage of the workpiece [2]. Taking the casing part of an aviation engine as an example, its processing state in the time domain [0, T] was set as $\{\boldsymbol{S}_t = \boldsymbol{S}(t) | 0 < t < T, \boldsymbol{S}_0 = \boldsymbol{S}(0)\}$, and its time-domain discrete method was shown in Fig. 2.4.

(1) The machining process from casing blank to finished product can be divided into rough machining, semi-finishing, finishing and polishing, etc. Therefore,

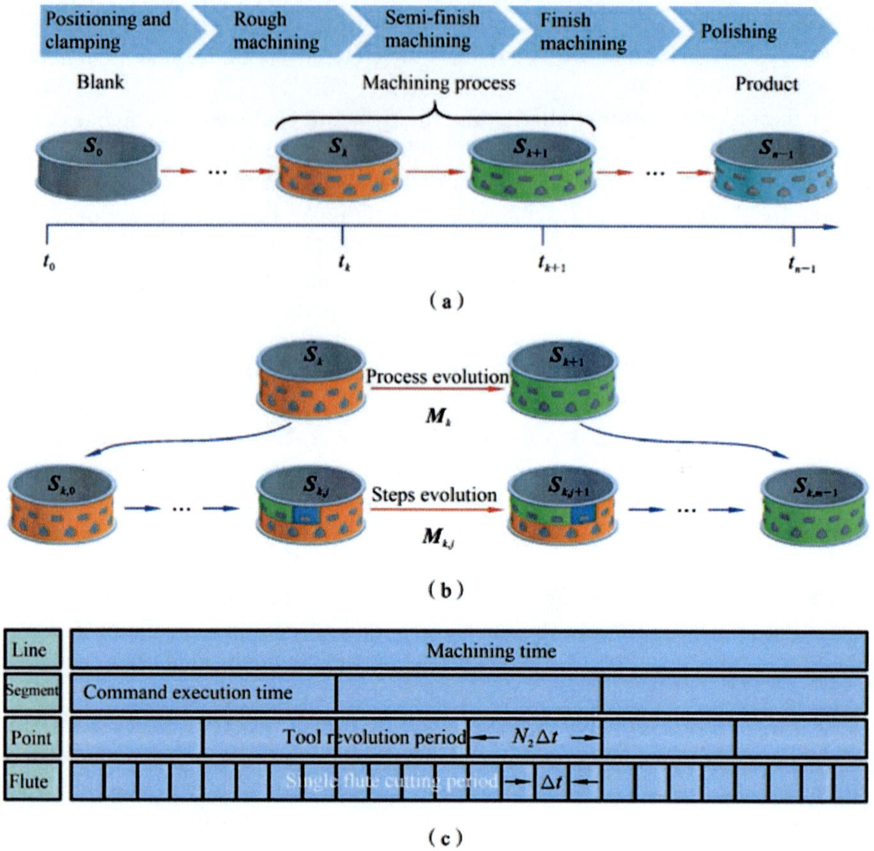

Fig. 2.4 Time-domain discrete method of the machining process. **a** discrete process based on process division **b** discrete machining process based on working step division **c** machining step dispersion based on trajectory division

the polymorphic process model can be defined as the state set $\{S_k = S(t_k)\}$ of $\{t_k | t_k < t_{k+1} (k = 0, 1, \ldots, n-1)\}$ at the beginning of each process.

(2) In addition, each procedure can be discretized into the work step state set $\{S_{k,j}\}$ on time series $\{t_{k,j} | t_{k,j} < t_{k,j+1} (j = 0, 1, \ldots, m-1)\}$ and the transfer subprocess set $\{M_{k,j}\}$ between adjacent work step states ($S_{k,j}$ and $S_{k,j+1}$) in the time domain $[t_k, t_{k+1}]$.

(3) The subprocess of transfer between adjacent working steps is defined by machining trajectory and cutting parameters and is realized by the margin cutting process between adjacent working steps. Among them, the machining trajectory is composed of several cutting line connections. The cutting line can be subdivided into the cutting segment and cutter location point, until the tooth cutting process. The existing geometric, mechanical and physical models

and time-domain discrete algorithms can be used to model and solve the cutting process in the cutting cycle.

2.2.3 Evolution of Polymorphic Models

In an intelligent programming system, the establishment of an accurate process model between each process is the key to realize intelligent processing technology, and researchers have done a lot of research on the establishment method of the process model. However, most of the current process model building methods are difficult to accurately and quickly obtain the process model of multiple states in the process from blank to finished product, that is, the polymorphic process model, so it cannot meet the processing requirements of intelligent processing with high efficiency.

The surface to be machined can be expressed as (u, v) parameters on a two-dimensional plane, where $0 \leq u, v \leq 1$. When there is a normal machining allowance, variable w can be added to represent the normal direction along the surface, thus forming a 3D mesh expression (u, v, w) of machining allowance. Any material in the margin to be removed corresponds to a unique position $\left(u_i, v_j, w_k\right)$ in the 3D grid. If the 3D mesh is segmented at a certain scale, each segmented 3D mesh will form a volume element. When the depth direction material is completely removed during the machining process, w coordinates can be ignored, and the volume element becomes the depth element [1]. The volume and depth elements do not need to strictly correspond to the geometry of the model but are only used to describe the material removal at the location determined, as shown in Figs. 2.5 and 2.6.

For each volume element mentioned above, the corresponding information model is defined. The information model consists of an information unit and data knowledge warehouse attached to the volume element. Each information unit is associated with the data knowledge warehouse through the volume element location coding to realize the process knowledge representation, storage and processing related to the volume element location. The data knowledge warehouse is used to record the process knowledge related to the location of the field data and volume elements.

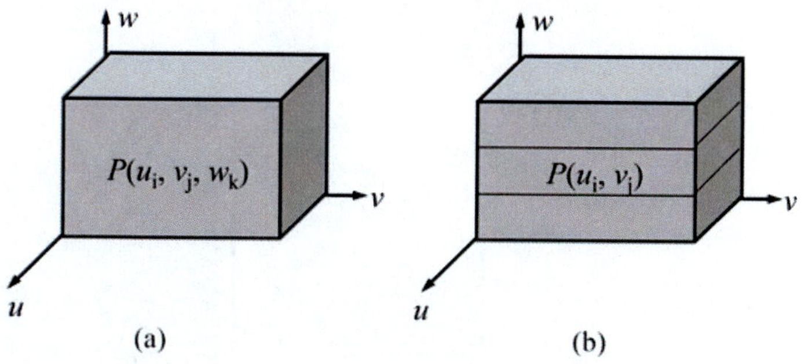

Fig. 2.5 Material volume element and depth element. **a** volume element **b** depth element

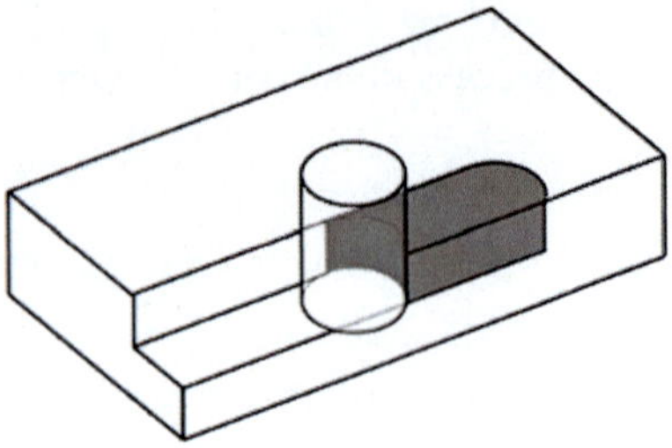

Fig. 2.6 Other expressions for volume element

The field data include the location of the volume element, processing signal and its time-position mapping relation; the process knowledge related to the location of the volume element includes the existing process knowledge and the knowledge obtained from the field data mining processing, such as the input vector $\boldsymbol{U}$, state vector $\boldsymbol{X}$, output vector $\boldsymbol{Y}$, identification results of model parameters $\boldsymbol{A}$ and $\boldsymbol{B}$ in local observation $\boldsymbol{Y} = \boldsymbol{AX} + \boldsymbol{BU}$, and dynamic response characteristic parameters such as local modal stiffness and modal frequency. The representation method of the state model $\boldsymbol{S}_k$ is shown in Fig. 2.7.

In the machining process, the processing location and processing state change constantly, and the state model information stored at each location also changes constantly. The process model is taken as an example to illustrate the polymorphic evolution in the machining process. Process model refers to the model of parts geometry, machining allowance, machining tools, cutting parameters, tool path, and process knowledge related to specific processing procedures in the process of machining, and it is related to the geometric shape and process requirements in the machining of part materials. The process model is polymorphic, which means that the process from blank to finished product can be divided into several states according

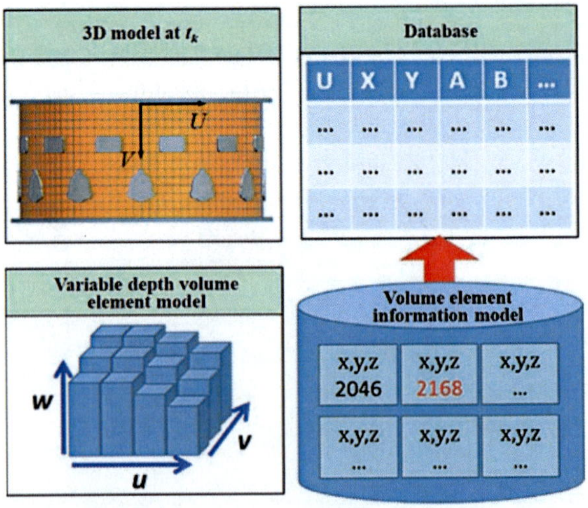

Fig. 2.7 Representation of the $\boldsymbol{S}_k$ state model

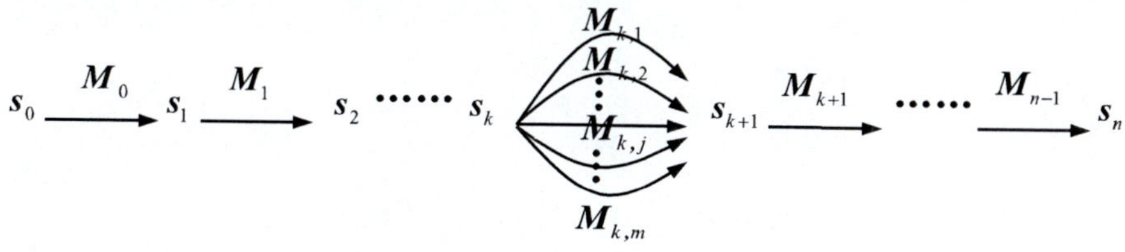

Fig. 2.8 Evolution of polymorphic process model

to the working step. A polymorphic process model can be defined as a set of each state process model, which can be expressed as:

$$\boldsymbol{S} = \{\boldsymbol{S}_k \mid k = 0, 1, 2, \ldots, n\} \tag{2.5}$$

where $\boldsymbol{S}$ is the set of polymorphic process models, $\boldsymbol{S}_k$ is the process model of state k, n is the number of working steps.

The change process from $\boldsymbol{S}_k$ to $\boldsymbol{S}_{k+1}$ in the polymorphic process model is called polymorphic process model evolution, and the evolution of polymorphic process model can be expressed as:

$$\boldsymbol{M}_k = \left\{\boldsymbol{M}_{k,j} \mid j = 0, 1, 2, \ldots, m\right\} \tag{2.6}$$

The evolution of the polymorphic process model is shown in Fig. 2.8, $\boldsymbol{S}_0 \to \boldsymbol{S}_1 \to \cdots \to \boldsymbol{S}_k \to \boldsymbol{S}_{k+1} \to \cdots \to \boldsymbol{S}_n$ represents the process model sequence of multiple states in the machining process. $\boldsymbol{M}_k$ represents the evolution model from the process model $\boldsymbol{S}_k$ to $\boldsymbol{S}_{k+1}$, and $\boldsymbol{M}_k$ has many evolutionary implementation paths. It is still a problem to define and optimize each evolutionary implementation path.

In the process of cutting, the process model evolution is realized by cutting margin, which involves margin distribution, tool sequence selection and the sequence of margin cutting.

2.3 Workpiece Geometric Evolution Model

The processing of complex parts is usually divided into several processing procedures, each procedure removes a certain amount of material, the shape of the workpiece is therefore constantly changing. In order to analyze the machining process more accurately, it is necessary to establish the geometric evolution model of the workpiece. The geometric evolution model of workpiece describes the change process of workpiece shape from blank to final part in rough machining, semi-finishing and finishing process. The establishment of this model is helpful to control the distribution of workpiece allowance between working procedures and provide model support for the programming of numerical control machining of complex parts in multiple working procedures.

2.3.1 Geometric Deformation Mapping Method

Geometric deformation mapping is a continuous, smooth and natural transition from one object (source object) to another (target object). Objects here can be digital images, polygons, polyhedral meshes, etc., or a special method of deformation geometry modeling [6]. This book will use deformation mapping technology to establish and express the process evolution model.

2.3.1.1 Representation of the Source and Target Objects

In the geometric deformation mapping technology, the shape of the source object and the target object directly control the shape of the intermediate deformation surface during deformation. Therefore, the representation of the source object and the target object is very important. If the transformation mapping technique is applied to the establishment of the transition surface family, there are two methods to represent the source object and the target object.

The first characterization method is to represent the source object with the transition blank before processing and the target object with the final shape of the finished product. When the workpiece and its blank part are determined, the boundary surface of the source object and the target object are determined. Figure 2.9a shows the schematic diagram of the source object and the target object under the first representation method of the impeller hub.

The second method is to represent the source object with an accurate surface boundary entity and the target object with a custom contraction center entity. Accurate surface boundary entity refers to simplified surface boundary entity derived from

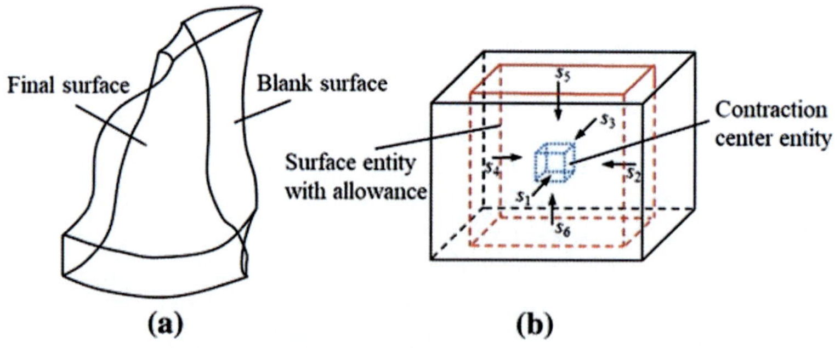

Fig. 2.9 Two kinds of representation of the source and target objects. **a** the first **b** the second

inside the cut geometry or semi-finished surface entity with non-uniform margin. In this way, the non-uniform margin distribution can be designed separately, so that the deformation process is not fertilized by the constraints of processing margin distribution requirements. If the boundary surface defined by the source object is continuously simplified within the allowance range of rough machining and semi-finishing, a simple contraction center entity can be defined in the source object as the target object of inward contraction and deformation of the source object. The deformation transition can be controlled by designing the solid shape of the object and the position of the shrinkage center. Figure 2.9b shows the schematic diagram of the source object and the target object of the cavity and groove part under the second representation method.

2.3.1.2 Parametric Method for 3D Entities

In NC machining, deformation mapping technology is realized by parameterizing the cut geometry which is determined by the Boolean difference between the source and target. It is necessary to describe the coordinates of each point inside the cut body in order to generate a smooth and varied intermediate deformation surface inside the cut body. The coordinate description of spatial geometry can be obtained through the coordinate transformation of a unit-body defined with ternary parameters and spatial geometry of the arbitrary shape. This transformation process is called the internal parameterization of three-dimensional entities.

When a three-dimensional entity of arbitrary shape is parameterized, its internal points can be represented by a set of ternary parameters, and the intermediate deformation surface can be obtained by fixing the parameters in a certain direction as constant values. This transformation needs to be realized by grid generation. Grid generation is a coordinate transformation process for mapping between the computational domain (u, v, w) that determines uniform orthogonal and the physical domain (x, y, z) that is non-uniform and non-orthogonal, as shown in Fig. 2.10, where $\boldsymbol{F}$ is the mapping transformation function.

In general, grid generation consists of two basic steps: first, to determine the distribution of discrete points on the boundary; the second is to determine the distribution of points within the boundary [7]. Similarly, three basic steps are required to parameterize the interior of a 3D entity. The first step is to determine the geometry to be removed in the machining process and six boundary surfaces. The second is to parameterize the six boundary surfaces of the cut geometry and determine the discrete points on the boundary surfaces. The third step is to determine the spatial points inside the cut geometry.

There are two main methods for parameterization of three-dimensional entities: one is a transfinite interpolation (TFI), which mainly applies the parameterization and interpolation methods and uses linear and nonlinear, one-dimensional or multidimensional interpolation formulas to generate grids; the other is the partial differential equation (PDE) method, which is mainly used for the generation of spatial surface

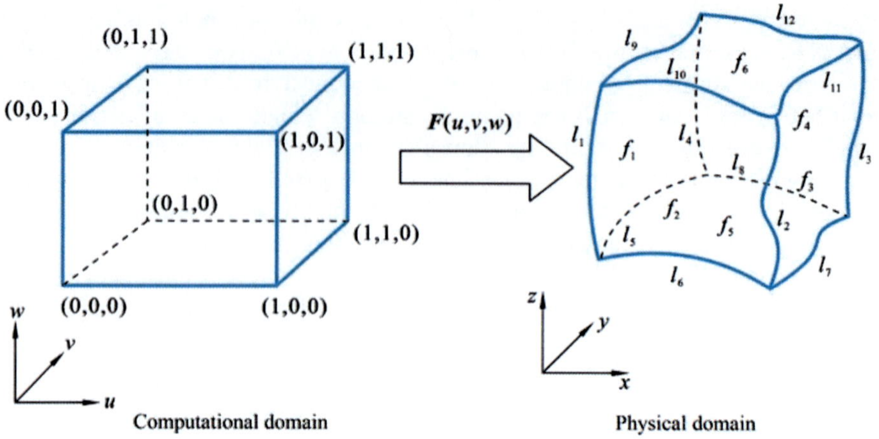

Fig. 2.10 The mapping transformation between the compute domain and the physical domain

meshes. By solving a set of partial differential equations between the computational domain and the physical domain, the meshes of the computational domain are transformed into the physical domain.

1. **Transfinite Interpolation Method**

The transfinite interpolation was proposed by William Gordon in 1973 [8]. In the early 1980s, Eriksson applied the transfinite interpolation to fluid dynamics to generate grids [9], and then improved on this basis and developed various forms of transfinite interpolation methods. At present, transfinite interpolation is still a widely used method for generating algebraic meshes. It can obtain meshes with specified boundaries by interpolation, and the spacing of meshes can be controlled directly.

The grid generation method of transfinite interpolation is to map the uniformly distributed and orthogonal computation point (u, v, w) in the computational domain to the non-uniformly distributed and non-orthogonal physical point (x, y, z) in the physical space through the mapping transformation function by determining an appropriate mapping. The mapping transformation process is shown in Fig. 2.11, and the mapping transformation function can be expressed as:

$$\boldsymbol{F}(u, v, w) = \begin{bmatrix} x(u, v, w) \\ y(u, v, w) \\ z(u, v, w) \end{bmatrix}, u, v, w \in [0, 1] \tag{2.7}$$

$\boldsymbol{F}$ is a vector-valued function, $\boldsymbol{F}(u_i, v_j, w_k)$ represents any structured grid, where:

$$\begin{cases} 0 \le u_i < \frac{i-1}{I-1} \le 1 \\ 0 \le v_j < \frac{j-1}{J-1} \le 1 \\ 0 \le w_k < \frac{k-1}{K-1} \le 1 \end{cases} \tag{2.8}$$

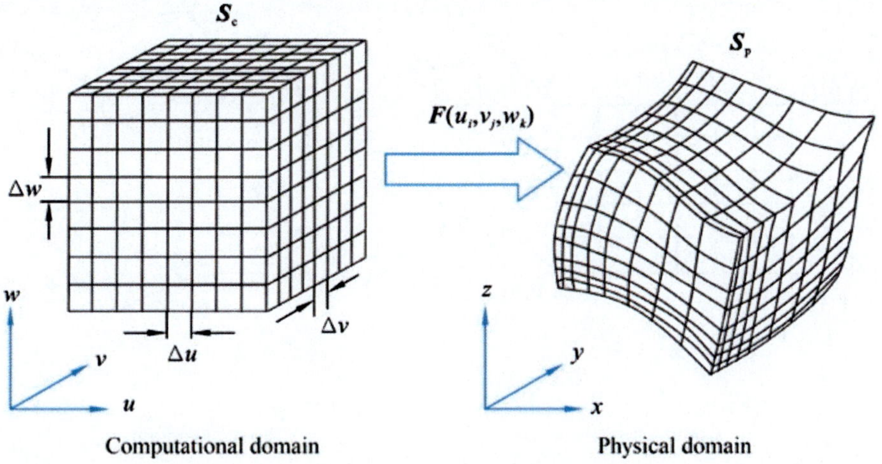

Fig. 2.11 Compute grid points in the domain and the physical domain

where $i = 1, 2, 3, \ldots, I$; $j = 1, 2, 3, \ldots, J$; $k = 1, 2, 3, \ldots, K$; I, J, K are grid numbers.

In the process of 3D entity parameterization, the boundary surfaces in the computational domain and physical domain are discretized first, and the corresponding boundary surfaces have the same number of discrete grid points, as shown in Fig. 2.11. Then, the uniformly discrete grid point (u_i, v_j, w_k) in the computational domain is transformed into the real grid point (x_i, y_j, z_k) in the physical domain by the mapping transformation function $\boldsymbol{F}(u, v, w)$. The mapping relationship between each neighboring grid point is constant and can be calculated by the mapping transformation function.

2. **Partial Differential Equation Method**

Partial differential equation methods mainly include the elliptic equation method, double curved equation method and parabolic equation method. Because the elliptic equation method is more suitable for the mesh generation of geometric entities in closed regions with specified physical boundaries, it can be well applied to the mesh generation of 3D geometric entities.

In partial differential grid generation, assuming that the grid distribution in the 3D entities in the physical domain is determined by a set of partial differential equations between the coordinates of the computational domain and the coordinates of the physical domain, and satisfying the requirements of boundary grid distribution, the grid in the computational domain can be transformed to the physical domain by solving this set of equations [10]. Therefore, for grid generation within 3D geometric entities, the core problem is to find the corresponding relationship between point (u, v, w) in the computational domain and point (x, y, z) in the physical domain, as shown in Fig. 2.12, where f is the boundary surface and l is the boundary curve.

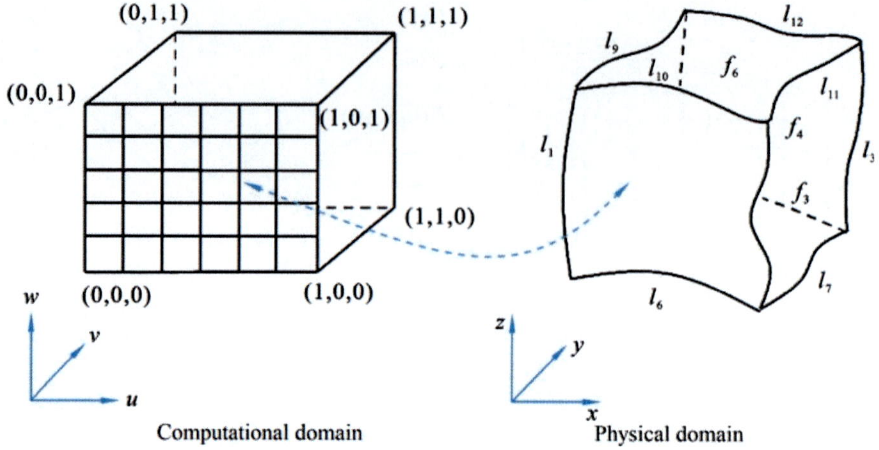

Fig. 2.12 A representation of the problem of generation grids for elliptic partial differential equations

If (x, y, z) and (u, v, w) are set as independent variables, this is a problem about finding the correspondence between variables (x, y, z) and (u, v, w) within the computational domain given the corresponding relations of them on the specific boundary. (This is equivalent to the first boundary condition. The first boundary condition is to find the distribution of variables at a given boundary).

From the physical domain, when u, v, w are the dependent variables of the unsolved solution in the physical domain, it will constitute a boundary value problem in the physical domain which is given the (u_B, v_B, w_B) corresponding to the boundary point (x_B, y_B, z_B) on the physical domain to find the (u, v, w) corresponding to the point (x, y, z) inside the physical domain. This correspondence can be mathematically described by partial differential equations with x, y, z as independent variables and u, v, w as dependent variables.

From the perspective of the computational domain, when x, y, z are regarded as the dependent variables to the unsolved solution in the computational domain, it will constitute a boundary value problem in the computational domain: specify the values of $x(u, v, w)$, $y(u, v, w)$, $z(u, v, w)$ on the rectangular boundary of the computational domain, and then solve the partial differential equation to determine the concrete value (x, y, z) in the physical domain, which means to find out the coordinates on the physical plane corresponding to each point in the computational plane solution region. In fact, when using elliptic partial differential equations, the generation of meshes is accomplished by solving boundary value problems in the computational domain. Therefore, it is necessary to transform the partial differential equation with x, y, z as the independent variables in the physical domain into the partial differential equation with u, v, w as the independent variablse in the computational domain.

The common solving steps of partial differential equations are as follows:

(1) Compute the governing equation.
(2) Compute the initial grid inside the physical space. The grid points obtained by the transfinite interpolation method can be used as the initial value of the initial grid without any special requirements on the quality of the initial grid. The initial grid point is treated as the first solution of the iterative algorithm, and the final grid quality is independent of the initial grid.
(3) Solve the partial differential equation system with the numerical or iterative method.

Because the partial differential equation system can generate smooth and uniformly distributed meshes when dealing with complex geometric shapes, it is often used as an internal meshes optimization method in practical applications.

2.3.1.3 An Internal Mesh Adjustment Method for Boundary Shrinkage

In the method of deformation mapping, the intermediate deformation surface can be constructed by defining three pairs of opposite boundary surfaces and generating the spatial mensh points by the mapping function. If the boundary surface is changed, the existing grid points can be constrained within the new boundary surface by the adjustment algorithm. The mesh point adjustment algorithm with a modified boundary is defined as an internal mesh adjustment method with a shrinking boundary. The calculation process of the algorithm is described in detail below.

Firstly, the parameter variable u_i, v_j, w_k in the calculation, the domain is replaced by i, j, k. Let any grid point in the physical domain be $\hat{F}(i, j, k)$, and the grid point on the boundary surface can be expressed as $F(i, j, k)$, $F(I, j, k)$, $F(i, 1, k)$, $F(1, J, k)$, $F(i, j, 1)$, and $F(i, j, K)$. Then, the adjusted grid $F'(i, j, k)$ can be expressed as:

$$\begin{cases} F_1(i,j,k) = F(i,j,k) + \alpha_1^0(u)\left[F(1,j,k) - \hat{F}(1,j,k)\right] + \alpha_2^0(u)\left[F(I,j,k) - \hat{F}(I,j,k)\right] \\ F_2(i,j,k) = F_1(i,j,k) + \beta_1^0(v)[F(i,1,k) - F_1(i,1,k)] + \beta_2^0(v)[F(i,J,k) - F_1(i,J,k)] \\ F'(i,j,k) = F_2(i,j,k) + \gamma_1^0(w)[F(i,j,1) - F_2(i,j,1)] + \gamma_2^0(w)[F(i,j,K) - F_2(i,j,K)] \end{cases} \tag{2.9}$$

where

$$\begin{cases} \alpha_1^0(u) = 1 - \xi_1(u) \quad \beta_1^0(v) = 1 - \eta_1(v) \quad \gamma_1^0(w) = 1 - \zeta_1(w) \\ \alpha_2^0(u) = \xi_2(u) \qquad \beta_2^0(v) = \eta_2(v) \qquad \gamma_2^0(w) = \zeta_2(w) \end{cases} \tag{2.10}$$

$$\begin{cases} \xi_1(u) = \frac{e^{C_1 u}-1}{e^{C_1}-1} \quad \eta_1(v) = \frac{e^{C_3 v}-1}{e^{C_3}-1} \quad \zeta_1(w) = \frac{e^{C_5 w}-1}{e^{C_5}-1} \\ \xi_2(u) = \frac{e^{C_2 u}-1}{e^{C_2}-1} \quad \eta_2(v) = \frac{e^{C_4 v}-1}{e^{C_4}-1} \quad \zeta_2(w) = \frac{e^{C_6 w}-1}{e^{C_6}-1} \end{cases} \tag{2.11}$$

C_1, C_2, C_3, C_4, C_5, C_6 are the influence parameters of the moving distance of the six boundary surfaces on the initial grid point, which is usually constant.

In the process of constructing the NC machining process model, the deformation mapping method is generally used to construct the intermediate deformation surface as the process surface along the w direction of cutting depth. However, in the actual machining process the geometric constraints will change with the change of the cutting depth, and the machining tool size corresponding to the process surface in the depth direction will also change, therefore, the boundary of the process surface needs to be modified to meet the actual machining requirements. It can be seen that the moving distance of the boundary surface is related to the size of the tool corresponding to the process surface in the depth direction, and C_1, C_2, C_3, C_4, C_5, C_6 can be set according to the moving distance.

It can be seen from the above description that determining the discrete points of the boundary surface is the key step of the deformation mapping method. At present, the discrete points are usually obtained by re-parameterizing the boundary surface, because the original parameter mesh of the boundary surface does not match the solid mesh, so the boundary surface must be re-parameterized according to the mesh division of the 3D entity. However, for most of the aviation parts with complex machining features, such as casing and impeller, the boundary features of the machining geometry usually include some complex surfaces with geometric constraints or irregular combined surfaces, which are difficult to be parameterized. In addition, in engineering applications, it is usually necessary to convert CAD models of parts into parametric defined surfaces, such as NURBS surfaces. Therefore, how to obtain the discrete points of complex boundary surface in CAD model becomes an urgent problem to be solved by the deformation mapping method. For the boundary definition of complex features, the constraints of complex boundaries are generally simplified and decomposed to reduce the difficulty of surface discretization, and then the simplified boundary surfaces are discretized to obtain the ordered discrete points of complex boundary surfaces in CAD models, so as to provide data support for subsequent algorithms.

2.3.2 *Deformation Mapping Modeling Method for Complex Machining Features*

In the process of deformation mapping modeling, through the generation and optimization of the inner mesh, the parameterization of the geometry to be machined can be realized, and then the polymorphic process model of the machined part can be obtained.

2.3.2.1 The Computational Flow of Deformation Mapping Modeling Method

The modeling process of complex machining feature deformation mapping is shown in Fig. 2.13. Firstly, analyze the complex characteristics of workpieces which contain the geometrical characteristics to determine its zoning method, the boundary surface discrete method and simplified method, the calculated results as the input of deformation mapping method, and then on the basis of the linear transfinite interpolation, use the Hermite transfinite interpolation and Lagrange transfinite interpolation to generate the cutting geometry within the grid. The Hermite transfinite interpolation can control the result of grid generation by controlling the length of normal vector mode, as Lagrange transfinite interpolation is by introducing virtual control surfaces. When the mesh surface satisfying the requirements cannot be obtained according to the over-limit interpolation results, the over-limit interpolation results are taken as the initial value of the PDE method, and the PDE method is used to optimize the distribution of the inner mesh. There are two kinds of partial differential equations commonly used: the Laplace partial differential equation and the Poisson partial differential equation. The Laplace partial differential equation can obtain mesh surfaces with smooth curvature, but lack of mesh screening, but the Poisson partial differential equation can be controlled by designing the governing equation for the inner mesh. Finally, a parametrized intermediate deformation surface can be obtained by fixing the parameter in a certain direction as constant.

The inner mesh distribution determines the geometric shape of the process surface and the margin distribution on the surface. However, the geometric shape and margin distribution of the process surface is closely related to the cutting process requirements. Therefore, in the process of deformation mapping modeling, the distribution of the inner mesh of the cut geometry must meet certain processing requirements.

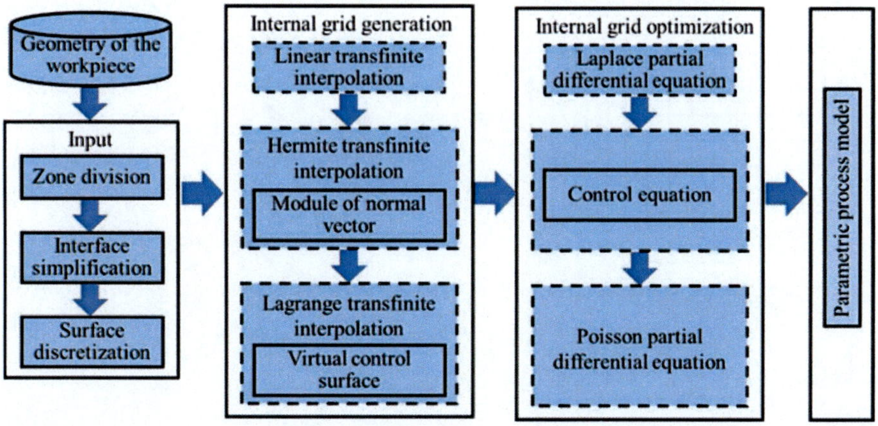

Fig. 2.13 Deformation mapping modeling process

2.3.2.2 The Internal Grid Generation Methods Based on Transfinite Interpolation

Since any single variable interpolation method can be applied to a single coordinate direction for interpolation calculation, different single variable interpolation methods or the combination of multiple single variable interpolation methods will produce countless different forms of transfinite interpolation methods. In general, higher-order or complex univariate interpolation methods are used to generate grids along the cutting direction, with the corresponding coordinate direction called the principal direction, while lower-order linear interpolation methods are used in the remaining two coordinate directions.

1. **Linear Transfinite Interpolation**

When the basis function of transfinite interpolation is linear, the method is called the linear transfinite interpolation method. The linear transfinite interpolation method is the simplest method for grid generation, as shown in Fig. 2.14. Assuming that the calculation domain, coordinate axes of the physical domain and the spatial positions of geometric information on the six surfaces of the cut body in the physical domain are determined, the linear interpolation basis function is:

$$\begin{cases} \alpha_1^0(u) = 1 - u & \beta_1^0(v) = 1 - v & \gamma_1^0(w) = 1 - w \\ \alpha_2^0(u) = u & \beta_2^0(v) = v & \gamma_2^0(w) = w \end{cases} \tag{2.12}$$

A univariate linear interpolation equation is constructed on each coordinate direction of the calculation domain, which is expressed as follows:

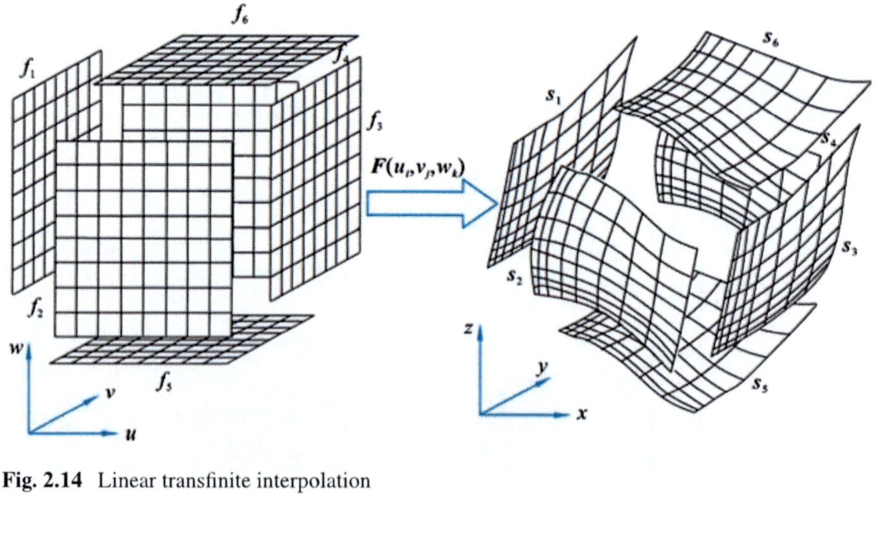

Fig. 2.14 Linear transfinite interpolation

$$\begin{cases} F_u(u, v, w) = (1-u)F(0, v, w) + uF(1, v, w) \\ F_v(u, v, w) = (1-v)F(u, 0, w) + vF(u, 1, w) \\ F_w(u, v, w) = (1-w)F(u, v, 0) + wF(u, v, 1) \end{cases} \tag{2.13}$$

The linear transfinite interpolation method is robust, fast and easy to calculate. However, this method does not allow much control over the intermediate deformation surface and is restrictively able to set the number of the intermediate deformation surface. Therefore, it is only suitable for shallow cavity whose shape changes little along the cavity depth direction.

2. **The Hermite Transfinite Interpolation Method**

The Hermite transfinite interpolation method combines the derivative value of the function to make the interpolation more accurate. If the basis function of cubic Hermite polynomial interpolation is taken as the basis function of the transfinite interpolation method, the transfinite interpolation is called cubic Hermite transfinite interpolation. The x, y, z coordinates in the physical domain correspond to the u, v, w coordinates in the computational domain, and the surface $s_1, s_2, s_3, s_4, s_5, s_6$ correspond to the plane $f_1, f_2, f_3, f_4, f_5, f_6$. In the transfinite interpolation method, if in the physical domain, the normal vector information of two corresponding points on the interpolation surface in a coordinate direction is known, the cubic Hermite transfinite interpolation method can be used in the coordinate direction. If z is the coordinate direction, then the univariate cubic Hermite interpolation ($N = 2$, $R = 1$) function in the z direction is:

$$\begin{aligned} F_w(u, v, w) &= \sum_{k=1}^{2}\sum_{l=0}^{1} \gamma_k^l \frac{\partial^l F(u, v, w_k)}{\partial w^l} \\ &= \gamma_0^1(w)(u, v, w_1) + \gamma_1^1(w)\frac{\partial F(u, v, w_1)}{\partial w} + \gamma_2^0(w)(u, v, w_2) \\ &+ \gamma_2^1(w)\frac{\partial F(u, v, w_2)}{\partial w} \end{aligned}$$

where $\gamma_1^0(w), \gamma_1^1(w), \gamma_2^0(w), \gamma_2^1(w)$ are interpolation basis functions, and their expressions are:

$$\begin{cases} \gamma_1^0(w) = \left(1 + 2\frac{w-w_1}{w_2-w_1}\right)\left(\frac{w-w_2}{w_1-w_2}\right)^2 \\ \gamma_2^0(w) = \left(1 + 2\frac{w-w_2}{w_1-w_2}\right)\left(\frac{w-w_1}{w_2-w_1}\right)^2 \\ \gamma_1^1(w) = (w-w_1)\left(\frac{w-w_2}{w_1-w_2}\right)^2 \\ \gamma_2^1(w) = (w-w_2)\left(\frac{w-w_1}{w_2-w_1}\right)^2 \end{cases} \tag{2.14}$$

where w_1, w_2 are the parameter values of two corresponding surfaces $F(u, v, w_1)$, $F(u, v, w_2)$ in the z direction.

Surfaces $F(u, v, w_1)$, $F(u, v, w_2)$ correspond to surfaces s_5, s_6. The partial derivatives of $F(u, v, w_1)$, $F(u, v, w_2)$ in the w direction can be determined by the cross product of the tangent vectors of the corresponding points on the surface in the u and v directions. The partial derivative in the w direction is going to be perpendicular to the surface $F(u, v, w_1)$, $F(u, v, w_2)$. Then, the partial derivative in the w direction is:

$$\begin{cases} \frac{\partial F(u,v,w_1)}{\partial w} = \left[\frac{\partial F(u,v,w_1)}{\partial u} \times \frac{\partial F(u,v,w_1)}{\partial v}\right] \psi_{w_1}(u, v) \\ \frac{\partial F(u,v,w_2)}{\partial w} = \left[\frac{\partial F(u,v,w_2)}{\partial u} \times \frac{\partial F(u,v,w_2)}{\partial v}\right] \psi_{w_2}(u, v) \end{cases} \tag{2.15}$$

where $\frac{\partial F(u,v,w_1)}{\partial u} \times \frac{\partial F(u,v,w_1)}{\partial v}$ is the direction of the partial derivative of surface $F(u, v, w_1)$ in the direction of w; $\frac{\partial F(u,v,w_2)}{\partial u} \times \frac{\partial F(u,v,w_2)}{\partial v}$ is the direction of the partial derivative of surface $F(u, v, w_2)$ in the direction of w; $\psi_{w_1}(u, v)$ and $\psi_{w_2}(u, v)$ respectively represent the length of partial derivatives of surface $F(u, v, w_1)$ and $F(u, v, w_2)$ in the w direction, and subscripts w_1 and w_2 are the corresponding parameter values of surface $F(u, v, w_1)$ and $F(u, v, w_2)$ in the depth direction.

In the Hermite transfinite interpolation method, the partial derivative (normal vector) of the interpolating surface provides additional control factors for the intermediate deformed surface. When the direction of the normal vector is determined, the modulus length of the normal vector becomes an important parameter to control the shape and interpolation accuracy of the intermediate deformation surface. If the norm length of the normal vector is defined as a constant or a scalar function related to surface information, the scalar function is the model parameter of the Hermite transfinite interpolation modeling method. Compared with the linear transfinite interpolation method, the Hermite interpolation method is more flexible and can control the shape of the intermediate deformation surface. The Hermite interpolation method is more suitable for a complex curved cavity or a deep cavity.

3. **The Lagrange Transfinite Interpolation Method**

The Lagrange transfinite interpolation is a polynomial interpolation method that can approximate complex curves according to given data points. If the basis function of Lagrange polynomial interpolation is taken as the basis function of the transfinite interpolation method, additional virtual control surfaces can be introduced into the deformation mapping modeling process to control the shape change of the process model. Such a transfinite interpolation method is called The Lagrange transfinite interpolation method.

As shown in Fig. 2.15, the x, y, z coordinates in the physical domain correspond to the u, v, w coordinates in the computational domain, and the surface $s_1, s_2, s_3, s_4, s_5, s_6$ correspond to the parametric surfaces $f_1, f_2, f_3, f_4, f_5, f_6$. In the transfinite interpolation method, if the position information of k corresponding surfaces in the computational domain and the physical domain is known (except for

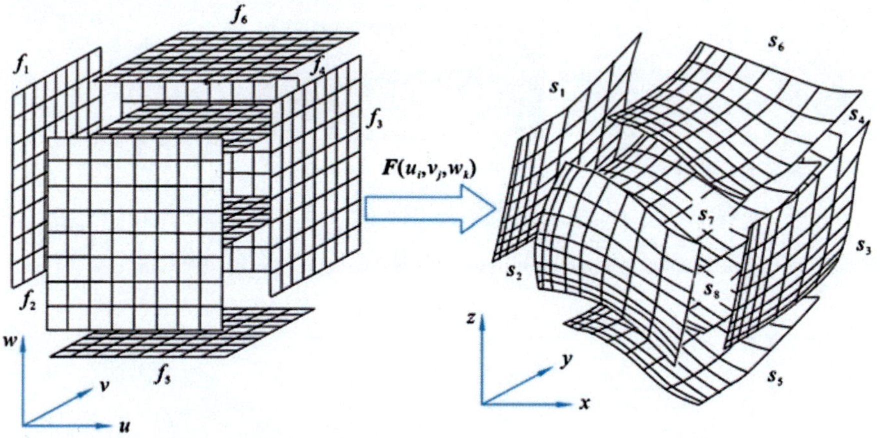

Fig. 2.15 The Lagrange transfinite interpolation

the six outer surfaces of the spatial geometry), the $m + 1$ Lagrange transfinite interpolation method can be used to generate the intermediate deformed surface. m is the number of virtual control surfaces, as shown in Fig. 2.15 when $m = 2$. It can be seen that the Lagrange transfinite interpolation method can introduce several additional virtual control surface constraints to the modeling process of deformation mapping. These virtual control surfaces are defined as the model parameters in the Lagrange transfinite interpolation method, and the model parameters are designed by setting the distribution position and geometric shape of the virtual control surfaces. Virtual control surfaces can be distributed in the interior, exterior or interior and exterior of the cut geometry. The specific situation depends on the geometric features of the parts. The number of virtual control surfaces is the same as the number of machining features of the parts.

2.3.2.3 Modeling Instance

This section takes impeller machining as an example to illustrate the application of the modeling method in the geometric state modeling of workpiece. For the impeller, the 70%~90% margin should be removed in the rough machining process of the impeller hub, so the calculation of the machining process model of the impeller hub is quite important. As shown in Fig. 2.16a, the impeller model has 11 blades, the height of the impeller is 25 mm, the inside diameter is 88.75 mm, and the outside diameter is 222.855 mm. The blade profile is a free-form curved surface with a maximum height of 39.35 mm. The impeller blank is a rotary part processed by a cylinder body through turning and milling, as shown in Fig. 2.16b. The cut geometry of the hub can be obtained by cutting the blank with two adjacent blades and hub surfaces of the impeller hub, as shown in Fig. 2.16c.

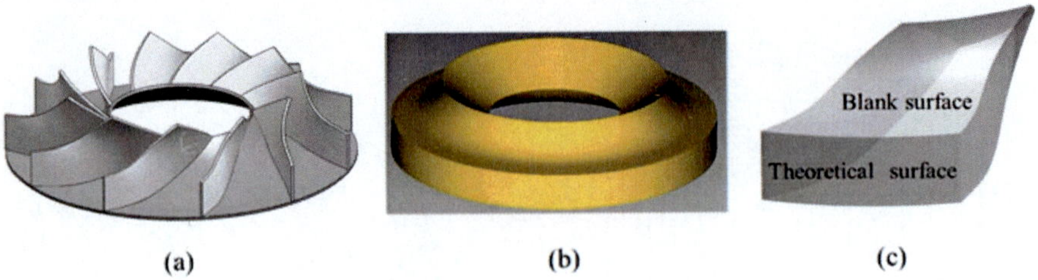

Fig. 2.16 Impeller model. **a** Impeller **b** Impeller Roughcast **c** Geometry to be cut

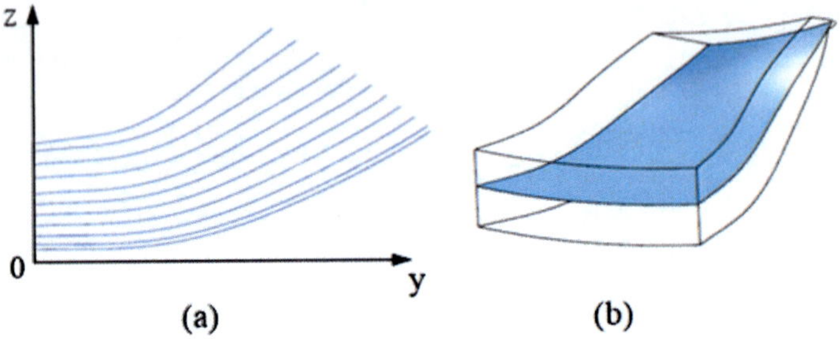

Fig. 2.17 Impeller channel process model. **a** the cross-section at u = 0.5 **b** the middle deformation surface of the flow passage

In Fig. 2.16c, the cut geometry of the hub is composed of six directly parameterizable boundary surfaces. The theoretical surface and the rough surface are discretized into 51 × 51 grid points, and the rest boundary surfaces are discretized into 51 × 11 grid points. After parameterization of the cut geometry by using the linear transfinite interpolation method, 9 intermediate deformation surfaces will be generated in their inner parts. These deformation surfaces are surface models that gradually transition from rough surfaces to theoretical surfaces, and their geometric shapes are similar to theoretical surfaces, as shown in Fig. 2.17. It can be seen from Fig. 2.17 that the intermediate deformation surface obtained by linear transfinite interpolation has relatively uniform spatial distribution and gentle curvature change, so it is suitable to be used as a procedure surface.

2.4 The Workpiece Dynamic Evolution Model

In the machining process of thin-walled parts, with the continuous cutting of the workpiece material, the tool direction and cutting position will constantly change, these phenomena will have a significant effect on the dynamic characteristics of the process system, making the cutting process of thin-walled parts show time-varying characteristics. With the continuous cutting of the workpiece material, the geometric structure of the workpiece is constantly evolving, and so do the dynamic characteristics of the corresponding workpiece subsystem.

2.4.1 Dynamic Evolution Analysis of Workpiece Based on Structural Dynamic Modification Technique

The dynamic structure modification can be used to predict the dynamic evolution of process system quickly and effectively [11]. In the cutting process of thin-walled parts, the modal parameters of the workpiece system will change continuously with the cutting process [2].

The dynamic equation of the process system can be expressed as:

$$\boldsymbol{M}\ddot{\boldsymbol{u}}(t) + \boldsymbol{C}\dot{\boldsymbol{u}}(t) + \boldsymbol{K}\boldsymbol{u}(t) = \boldsymbol{F}(t) \tag{2.16}$$

where: $\boldsymbol{M}$ is the modal mass matrix; $\boldsymbol{C}$ is the modal damping matrix; $\boldsymbol{K}$ is the modal stiffness matrix; $\boldsymbol{u}(t)$ is the displacement vector; $\boldsymbol{F}(t)$ is the cutting force vector.

When the cutting force is the harmonic force whose excitation frequency is w, the corresponding system displacement is:

$$\boldsymbol{u}(t) = \left(-\omega^2\boldsymbol{M} + i\boldsymbol{C} + \boldsymbol{K}\right)^{-1}\boldsymbol{F}(t) = \boldsymbol{A}\boldsymbol{F}(t) \tag{2.17}$$

where $\boldsymbol{A}$ is the dynamic flexibility matrix of the machining system.

According to the theory of structural dynamic modification, the dynamic matrix of the structural modification part, namely the removed material, is:

$$\boldsymbol{D} = -\omega^2\Delta\boldsymbol{M} + i\Delta\boldsymbol{C} + \Delta\boldsymbol{K} \tag{2.18}$$

where $\Delta\boldsymbol{M}$ is the modal mass matrix of the modified part of the structure; $\Delta\boldsymbol{C}$ is the modal damping matrix of the modified part of the structure; $\Delta\boldsymbol{K}$ is the modal stiffness matrix of the modified part of the structure.

After the modification, the dynamic flexibility matrix of the process system is:

$$\boldsymbol{B} = \left[-\omega^2(\boldsymbol{M} + \Delta\boldsymbol{M}) + i(\boldsymbol{C} + \Delta\boldsymbol{C}) + (\boldsymbol{K} + \Delta\boldsymbol{K})\right]^{-1} \tag{2.19}$$

Therefore, the dynamic flexibility matrix of the process system before and after the structural dynamic modification has the following relationship:

$$\boldsymbol{B}^{-1} = \boldsymbol{A}^{-1} + \boldsymbol{D} \tag{2.20}$$

By transforming the above equation, yields:

$$\boldsymbol{B} = [\boldsymbol{I} + \boldsymbol{AD}]^{-1}\boldsymbol{A} \tag{2.21}$$

where $\boldsymbol{I}$ is the identity matrix.

Taking aeroengine thin-walled casing part as an example, the dynamic characteristic change of the process system caused by material removal in the cutting process of the casing can be regarded as the local structural dynamic modification of the process system, namely:

$$\boldsymbol{D} = \begin{bmatrix} \boldsymbol{D}_{11} & \boldsymbol{0} \\ \boldsymbol{0} & \boldsymbol{0} \end{bmatrix} \tag{2.22}$$

Substitute Eq. (2.22) into Eq. (2.21), yields:

$$\begin{bmatrix} \boldsymbol{B}_{11} & \boldsymbol{B}_{12} \\ \boldsymbol{B}_{21} & \boldsymbol{B}_{22} \end{bmatrix} = \left\{ \boldsymbol{I} + \begin{bmatrix} \boldsymbol{A}_{11} & \boldsymbol{A}_{12} \\ \boldsymbol{A}_{21} & \boldsymbol{A}_{22} \end{bmatrix} \begin{bmatrix} \boldsymbol{D}_{11} & \boldsymbol{0} \\ \boldsymbol{0} & \boldsymbol{0} \end{bmatrix} \right\}^{-1} \begin{bmatrix} \boldsymbol{A}_{11} & \boldsymbol{A}_{12} \\ \boldsymbol{A}_{21} & \boldsymbol{A}_{22} \end{bmatrix} \tag{2.23}$$

The solution process of structural dynamic modification does not consider the increase or decrease of the degree of freedom of the process system, which is feasible for the semi-finishing and finishing process of thin-walled parts. Because of the small volume of the material removal, especially the structural changes in the normal direction of the casing surface are not obvious. But for the rough machining of this kind of parts, if the increase or decrease of the degree of freedom of the process system is not considered, the prediction accuracy will be seriously reduced.

It can be seen from Eq. (2.23) that when the initial modal parameters of the process system and the modal parameters of the structural modification part are known, it is easy to calculate the modal parameters after the structural modification of the process system. The modal parameters of the initial and modified parts of the process system can be calculated by the finite element method.

2.4.2 *Dynamic Evolution Analysis of Workpieces Based on the Thin Shell Model*

2.4.2.1 A Common Hyperbolic Thin Shell Mechanical Model of the Shell Structure

Shell structures are common in aerospace equipment structures. The geometric features of these casing can be simplified to some extent by characterizing spherical surfaces, double parabolic surfaces, cylindrical surfaces, and other complex shapes, as shown in Fig. 2.18. Curved elements can better simulate real structures, and the results will be more efficient. However, there is a difference between the deformation of the surface shell and the deformation of the plate. For curved elementsthat consider lateral shear deformation are often used. As can be learned from Fig. 2.18,

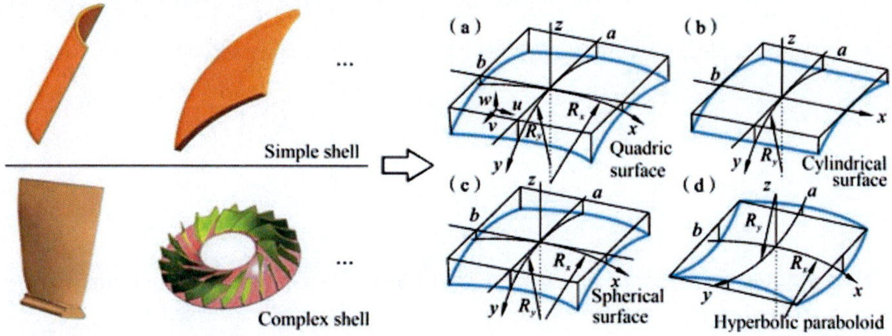

Fig. 2.18 Common shell structures

shell structures generally have different or identical radiuses in one or both directions, in addition to distortions in one direction. Compared to the plate, the approximate result of the double-curved thin shell is more accurate, so the double-curved thin shell is introduced to approximate the common shell [12]. This book mainly introduces theis presented as a double-curved thin shell.

Consider a double-curved shallow shell in which a surface can be represented in an orthogonal absolute coordinate system as an equation:

$$z = \frac{x^2}{2\mathrm{R}_x} + \frac{xy}{\mathrm{R}_{xy}} + \frac{y^2}{2\mathrm{R}_y} \tag{2.24}$$

where R_x and R_y are curvature radius in the x and y directions, R_{xy} is the warp radius of the surface and represents the distortion degree of the surface.

For analytical convenience, R_x, R_y, and R_{xy} are usually defined as constants, in which case the Eq. (2.24) represents a quadric surface, as shown in Fig. 2.18a; when $R_y = R$ and $R_x = R_{xy} = \infty$, the Eq. (2.24) represents the cylindrical surface, as shown in Fig. 2.18b; and when $R_x = R_y = R$ and $R_{xy} = \infty$, the Eq. (2.24) represents the spherical surface, as shown in Fig. 2.18c; when $R_x = -R_y = R$ and $R_{xy} = \infty$, the Eq. (2.24) represents the hyperbolic paraboloid, as shown in Fig. 2.18d. The neutral face of the shell is a surface, shown in Fig. 2.19. Assuming that the middle

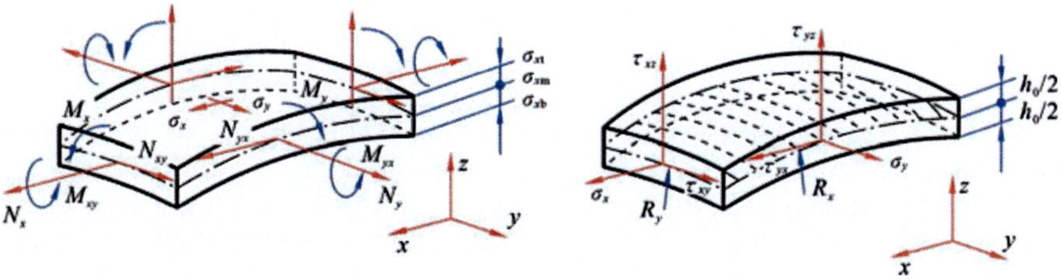

Fig. 2.19 The force and stress state of shell elements

surface of the shell meets Kirchhoff's hypothesis, the strain ε_0, torsional strain γ_0, and curvature change κ of the medium surface can be expressed as:

$$\begin{cases} \varepsilon_{0x} = \frac{\partial u}{\partial x} + \frac{w}{R_x}, \varepsilon_{0y} = \frac{\partial v}{\partial y} + \frac{w}{R_y} \\ \gamma_{0xy} = \frac{\partial v}{\partial x} + \frac{\partial u}{\partial y} + \frac{2w}{R_{xy}} \\ \kappa_x = -\frac{\partial^2 w}{\partial x^2}, \kappa_y = -\frac{\partial^2 w}{\partial y^2}, \kappa_{xy} = -\frac{\partial^2 w}{\partial x \partial y} \end{cases} \tag{2.25}$$

where u, v and w are displacements in the x, y and z directions of the medium surface. Strain ε and torsional strain γ at any point in the shell are expressed as:

$$\begin{cases} \varepsilon_x = \varepsilon_{0x} + z\kappa_x \\ \varepsilon_y = \varepsilon_{0y} + z\kappa_y \\ \gamma_{xy} = \gamma_{0xy} + 2z\kappa_{xy} \end{cases}$$

A micro-element is taken from the shell for analysis and the combined forces N_x, N_y and N_{xy} on the unit length are tangent with the neutral surface. The force balance equations of the lateral shear force can be selected, as shown in Fig. 2.19, for the combined force moments M_x, M_y and M_{xy} in unit length:

$$\begin{cases} \frac{\partial N_x}{\partial x} + \frac{\partial N_{xy}}{\partial y} + p_x = -\rho \frac{\partial^2 u}{\partial t^2} \\ \frac{\partial N_y}{\partial y} + \frac{\partial N_{xy}}{\partial x} + p_y = -\rho \frac{\partial^2 v}{\partial t^2} \\ -\left(\frac{2N_{xy}}{R_{xy}} + \frac{N_x}{R_x} + \frac{N_y}{R_y}\right) + \frac{\partial^2 M_x}{\partial x^2} + \frac{\partial^2 M_{xy}}{\partial y \partial x} + \frac{\partial^2 M_y}{\partial y^2} + p_n = -\rho \frac{\partial^2 w}{\partial t^2} \end{cases} \tag{2.26}$$

where p_x and p_y are the forces of external forces in the direction of cutting, p_n is the force in the direction of the law, and ρ is the density of materials. For an isotropic shell material, the relationship between stress and strain is:

$$\varepsilon_x = \frac{1}{E}\left(\sigma_x - \nu\sigma_y\right), \varepsilon_y = \frac{1}{E}\left(\sigma_y - \nu\sigma_x\right), \gamma_{xy} = \frac{\tau_{xy}}{G} \tag{2.27}$$

$$\sigma_x = \frac{E}{1-\nu^2}\left(\varepsilon_x + \nu\varepsilon_y\right), \sigma_y = \frac{E}{1-\nu^2}\left(\varepsilon_y + \nu\varepsilon_x\right) \tag{2.28}$$

$$\tau_{xy} = \frac{E}{2(1+\nu)}\gamma_{xy} \tag{2.29}$$

where σ is stress; τ is torsional stress; E is Elastic mod (Yang's mod); ν is Poisson ratio; G is the shear mode, which meets $G = E/[2(1+\nu)]$. Integration of the corresponding force on the section can find a joint force, and integration of the corresponding force on the thickness h_0 can obtain a bending moment:

$$\begin{cases} N_x = \int_{-h_0/2}^{h_0/2} \sigma_x \mathrm{dz} \\ N_y = \int_{-h_0/2}^{h_0/2} \sigma_y \mathrm{dz} \\ N_{xy} = \int_{-h_0/2}^{h_0/2} \tau_{xy} \mathrm{dz} \\ M_x = \int_{-h_0/2}^{h_0/2} \sigma_x \mathrm{zdz} \\ M_y = \int_{-h_0/2}^{h_0/2} \sigma_y \mathrm{zdz} \\ M_{xy} = \int_{-h_0/2}^{h_0/2} \tau_{xy} \mathrm{zdz} \end{cases} \tag{2.30}$$

The joint moment is represented as the:

$$\begin{cases} M_x = D(\kappa_x + \nu\kappa_y) \\ M_y = D(\kappa_y + \nu\kappa_x) \\ M_{xy} = D(1-\nu)\kappa_{xy} \end{cases} \tag{2.31}$$

where

$$D = \frac{Eh_0^3}{12(1-\nu^2)}$$

where D is the anti-bend stiffness of the shell. Under the orthogonal coordinate system, the motion equation of the shallow shell is expressed as a displacement matrix as:

$$\begin{bmatrix} -p_x \\ -p_y \\ p_n \end{bmatrix} = \frac{Eh_0}{1-\nu^2} \begin{bmatrix} \ell_{11} & \ell_{12} & \ell_{13} \\ \ell_{21} & \ell_{22} & \ell_{23} \\ \ell_{31} & \ell_{32} & \ell_{33} \end{bmatrix} \begin{bmatrix} u \\ v \\ w \end{bmatrix} + \rho h_0 \begin{bmatrix} -\frac{\partial^2 u}{\partial t^2} \\ -\frac{\partial^2 v}{\partial t^2} \\ -\frac{\partial^2 w}{\partial t^2} \end{bmatrix} \tag{2.32}$$

where $l_{ij}(i, j = 1, 2, 3)$ is the differential operator.

According to the classical Ritz energy method, potential energy in the shell is produced by strain energy, expressed as:

$$PE_m = PE_s + PE_b \tag{2.33}$$

where PE_s is produced by the extruded intermediate surface, PE_b is produced by the bending moment [13], and:

$$PE_s = \frac{1}{2}\int\int \frac{Eh_0^2}{(1-\nu^2)} \left\{ (\varepsilon_{0x} + \varepsilon_{0y})^2 - 2(1-4\nu)\left(\varepsilon_{0x}\varepsilon_{0y} - \frac{\gamma_{0xy}^2}{4} \right) \right\} dA_s \tag{2.34}$$

where A_s is the area of the surface in the shell.

In the shell, bending produces the potential for:

$$PE_b = \frac{1}{2}\int\int D\Big((\kappa_x + \kappa_y)^2 - 2(1-\nu)\big(\kappa_x\kappa_y - \kappa_{xy}^2\big)\Big)dA_s \tag{2.35}$$

The kinetic energy of vibration is:

$$KE_m = \frac{1}{2}\int\int \rho h_0\left(\frac{\partial u}{\partial t}\right)^2\left(\frac{\partial v}{\partial t}\right)^2\left(\frac{\partial w}{\partial t}\right)^2 dA_s \tag{2.36}$$

from Eq. (2.34) to Eq. (2.36), the energy expression is obtained by the function integral containing Young's modulus E, material density ρ, shell thickness h_0, so the above-mentioned method of solving kinetic and potential energy is suitable for shells of variable thickness or non-homogeneous materials.

For a shell with a fixed clip at the bottom, it needs to meet zero displacement in x, y, z direction and zero speed in z direction, which means to satisfy Eq. (2.37) to Eq. (2.41):

$$N_{0x} - N_x = 0 \text{ or } u_0 = 0 \tag{2.37}$$

$$\left(N_{0y} - \frac{M_{0xy}}{R_y}\right) - \left(N_{xy} - \frac{M_{xy}}{R_y}\right) = 0 \text{ or } v_0 = 0 \tag{2.38}$$

$$\left(Q_{0x} - \frac{\partial M_{0xy}}{\partial y}\right) - \left(Q_x - \frac{\partial M_{xy}}{\partial y}\right) = 0 \text{ or } w_0 = 0 \tag{2.39}$$

$$M_{0x} - M_x = 0 \text{ or } \frac{\partial w_0}{\partial x} = 0 \tag{2.40}$$

$$M_{0xy} w_0 \big|_{y_1}^{y_0} = 0 \tag{2.41}$$

2.4.2.2 Modeling of Cutting Depth Under Different Cutting Schemes

In the process of shallow shell machining, if different machining schemes are used, the part shape will be different in the inter-machining stage, but the main difference is that the cutting depth changes differently. In general, there are three main machining schemes: one-way cutting, re-cutting, and surround cutting.

Projecting the shallow shell onto the Oxy plane creates a closed area. To keep the analysis simple, the x-axis and the y-axis are redefined in a different orthogonal absolute coordinate system, as shown in Fig. 2.20. In order to simulate the cutting process, the projection area of the shallow shell is divided into nine sub-regions. From left to right, the boundary curve and the split curve are represented

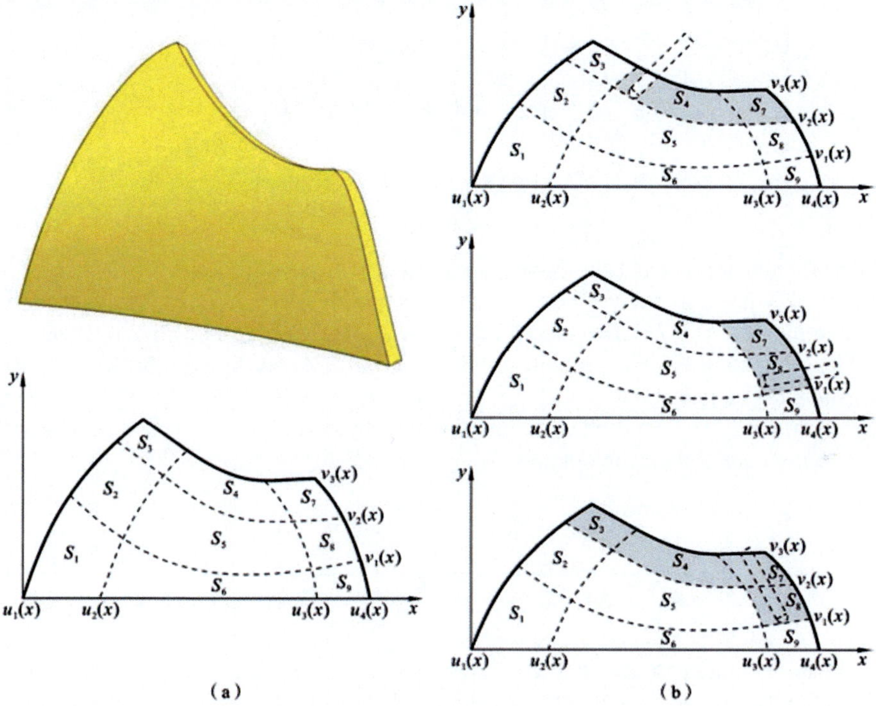

Fig. 2.20 Typical shell structure and its projection and removal sequence on the Oxy plane. **a** typical projection of the structure on the Oxy plane **b** typical removal sequence

by $u_1(x)$, $u_2(x)$, $u_3(x)$, $u_4(x)$, respectively. From the bottom up, the boundary curve and the split curve are represented by $v_1(x)$, $v_2(x)$, $v_3(x)$ in turn. The nine subregions are marked S_1, S_2, S_3, S_4, S_5, S_6, S_7, S_8, S_9. The thickness of each subregion is H_1, H_2, H_3, H_4, H_5, H_6, H_7, H_8, H_9. Using the paramethy model above, the corresponding segment smooth curve and the thickness of each area can be adjusted to simulate the machining process. Figure 2.20b shows some of the possible conditions during the cutting process, where the gray area represents the machined area and its thickness changes.

2.4.2.3 Modal Frequency and Modal Vibration Solution

This book uses the Riley-Ritz method to solve the natural frequency and modal vibration pattern of the thin shell at any stage of the machining process. The vibration equation of the thin shell assumes that, in order to satisfy the sine function of the boundary constraints of the cantilever beam, it can be represented by the following equation:

$$\begin{cases} u(x,y,t) = \sum_{i=0}^{I_c}\sum_{j=0}^{J_c} \alpha_{i,j}\left(\frac{x}{a}\right)^i\left(\frac{y}{b}\right)^j \sin\omega t = U(\varepsilon,\eta)\sin\omega t \\ v(x,y,t) = \sum_{k=0}^{K_c}\sum_{l=0}^{L_c} \beta_{k,l}\left(\frac{x}{a}\right)^k\left(\frac{y}{b}\right)^l \sin\omega t = V(\varepsilon,\eta)\sin\omega t \\ w(x,y,t) = \sum_{m=0}^{M_c}\sum_{n=0}^{N_c} \gamma_{m,n}\left(\frac{x}{a}\right)^m\left(\frac{y}{b}\right)^n \sin\omega t = W(\varepsilon,\eta)\sin\omega t \end{cases} \tag{2.42}$$

where a and b represent the corresponding horizontal width and vertical height of the shell, while ε and η are the dimensionless coordinates corresponding to x and y, which means $\varepsilon = x/a$, $\eta = y/b$; I_c, J_c, K_c, L_c, M_c, N_c are all positive integer.

In order to solve the problem of free vibration, according to the Ritz energy method, the displacement equation is brought into the energy Eq. (2.33) and the total potential energy of the shell is minimized relative to coefficients $\alpha_{i,j}$, $\beta_{k,l}$, $\gamma_{m,n}$, the following equation can be obtained:

$$\begin{cases} \frac{\partial(KE_m - PE_m)}{\partial\alpha_{i,j}} = 0, i = 0,1,2,\ldots,I_c;\ j = 0,1,2,\ldots,J_c \\ \frac{\partial(KE_m - PE_m)}{\partial\beta_{k,l}} = 0, k = 0,1,2,\ldots,K_c;\ l = 0,1,2,\ldots,L_c \\ \frac{\partial(KE_m - PE_m)}{\partial\gamma_{m,n}} = 0, m = 0,1,2,\ldots,M_c;\ n = 0,1,2,\ldots,N_c \end{cases} \tag{2.43}$$

Feature values can be obtained by solving the corresponding equation. The corresponding feature vector can then be calculated by bringing the feature value into Eq. (2.43). Finally, the eigenvector is brought into Eq. (2.42), and the modal vibration pattern corresponding to each frequency can be obtained.

As can be learned from the above analysis, the complexity of the resolution method depends on the complexity of the model. When the contour shape of the shell or the geometry of the intermediate stage of the machining process is very complex, the calculation of the analysis method is too large to be achieved. Therefore, it is often necessary to simplify the model, such as approximate the boundary curve to a straight line or analyze only a few specific intermediate stages.

2.5 Tool Wear Evolution Model

2.5.1 Tool Wear During the Machining Process

Among the two subsystems introduced in Sect. 2.1, the most time-varying factor in the tool-spindle subsystem is the tool wear. Tool flank wear is a strong time-varying factor in the cutting process of aviation difficult-to-cut material. When the amount of tool wear reaches a certain limit, it will seriously affect the thermal coupling of cutting forces and the quality of the machined surface, which will affect the performance of key aviation components during service. Therefore, the establishment of a tool wear evolution model and real-time monitoring of tool wear status during machining is extremely important to ensure the machining quality of key aviation components. In the cutting process, common tool wear types and their corresponding wear mechanisms are shown in Table 2.1, and common tool wear forms are shown in Fig. 2.21. The most common and most studied type of tool wear is flank wear. The common tool life curve based on flank tool wear is shown in Fig. 2.22. Tool wear can generally be divided into three stages: initial wear stage, normal wear stage and accelerated wear stage. In the cutting of aerospace materials, the wear when using cemented carbide tools to cut titanium alloys is generally formed by the combined

Table 2.1 Common types of tool wear and their mechanisms

Wear type	Mechanism	Schematic diagram
Abrasive wear	Hard particles in the material scratch the surface of the tool, causing tool wear. This is also one of the main factors causing tool flank wear	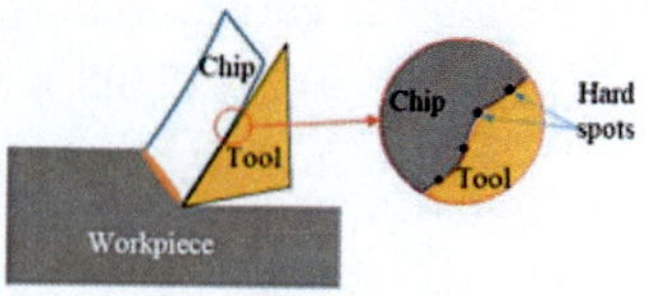
Adhesive wear	Under the action of high temperature, high pressure and friction, the chips stick to the surface of the tool, which is easy to cause the built-up edge of the tool and cause the shape of the tool edge to change or the groove to appear	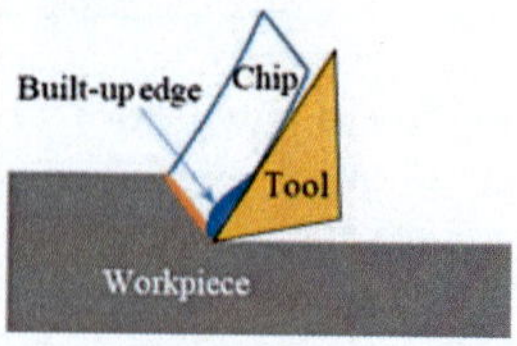
Diffusion wear	At high temperatures, the tool material atoms diffuse into the chips or the workpiece, thereby weakening the strength of the cutting edge of the tool, causing chipping or breakage	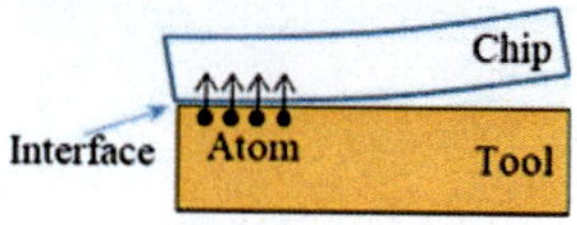
Oxidation wear	Oxidation wear is caused by the oxidation of tool materials in the air environment	–
Thermal wear	On the one hand, the high temperature in the cutting process will cause the softening of the tool material and plastic deformation of the tool; on the other hand, frequent heating–cooling cycles will cause thermal cracks on the tool	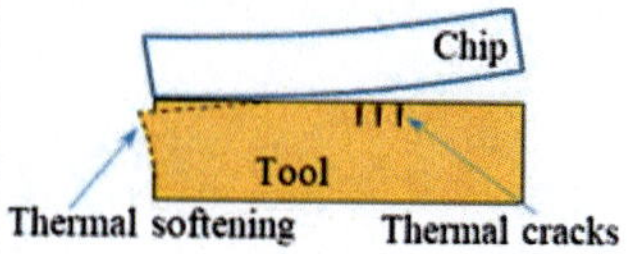

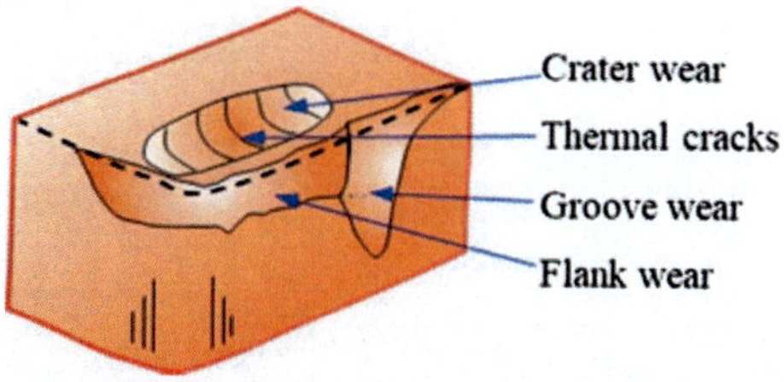

Fig. 2.21 Common tool wear forms

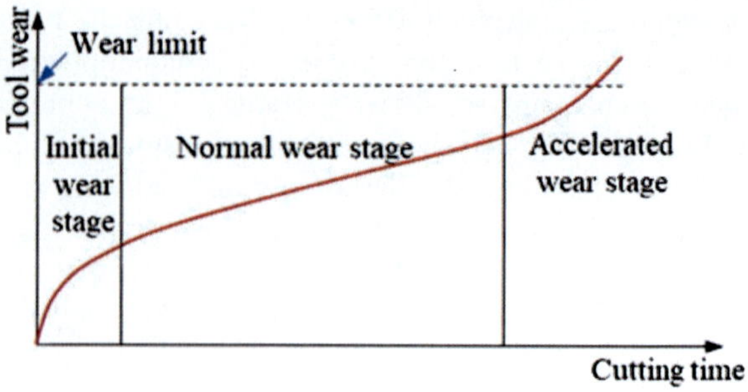

Fig. 2.22 The tool life curve

effects of abrasive wear, adhesive wear and diffusion wear; when using cemented carbide tools to cut super-alloy, the main wear of the tool types are abrasive wear, diffusion wear and chipping.

2.5.2 *Evolutionary Modeling of Tool Flank Wear*

According to the law of tool wear in stages, the wear amount and wear speed of the tool flank are used as state parameters to establish a prediction model of tool wear evolution [14]:

$$VB(t_k) = VB(t_{k-1}) + \int_{t_{k-1}}^{t_k} r(t)dt = VB(t_{k-1}) + r_k(t_k - t_{k-1}) + \frac{1}{2}\lambda_k(t_k - t_{k-1})^2 \tag{2.44}$$

where: $t_{k-1} \le t \le t_k$, $VB_{(t_k)}$ is the total amount of tool wear at time t_k, r_k is the tool wear rate at time t_k, λ_k is the tool wear acceleration at time t_k.

In general, according to the three-stage wear law, Eq. (2.44) has different expressions in the three stages, as shown in Fig. 2.23.

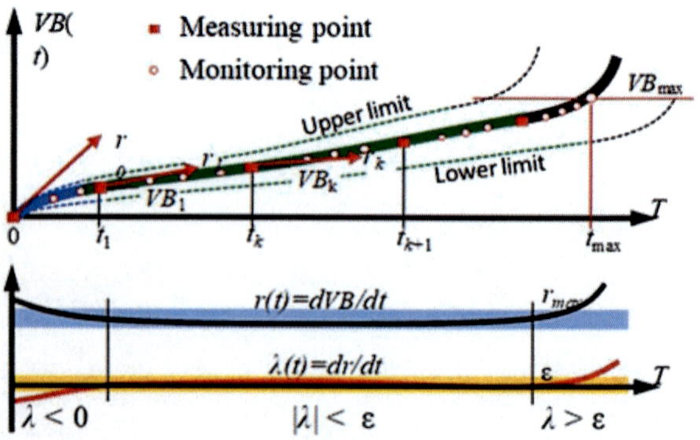

Fig. 2.23 Schematic diagram of tool wear monitoring and control

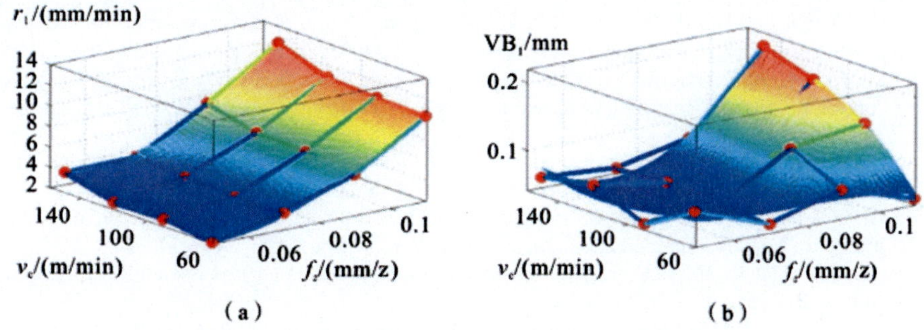

Fig. 2.24 Tool wear experimental modeling. **a** tool wear rate **b** flank wear

(1) Initial wear stage: the tool wear at this stage is a quadratic curve of time, which can be expressed as:

$$VB_0(t) = VB_0 + r_0 t + \frac{r_1 - r^*}{t_1} t^2 \tag{2.45}$$

(2) Normal wear stage: Under normal circumstances, the tool wear in this stage has a linear relationship with time, which can be expressed as:

$$VB_k(t) = VB_k + r_k(t - t_k), k \geq 1 \tag{2.46}$$

During the machining process, the detection and identification monitoring points of tool wear are set. Through monitoring and calculation, the tool wear amount VB_k and wear rate r_k at time t_k can be obtained, and the prediction of the tool wear amount $VB_k(t)$ when t > t_k is realized by Eq. (2.46).

(3) Rapid wear stage: When the detection result shows $r_k > r_{\max}$ or $\lambda_k > \varepsilon$ or $VB_k \geq VB_{\max}$, the tool enters the rapid wear state and triggers the stop processing and tool replacement instructions.

Under certain machining conditions, where r_k and VB_k are functions of cutting speed v_c and feed per tooth f_z, they can be modeled by experimental methods. Figure 2.24a, b show the relationship between the tool wear rate r_1 and the flank wear VB_1 at t_1 obtained through experiments with the cutting speed v_c and the feed per tooth f_z.

References

1. LUO M. Dynamics modeling and machining process control of low rigidity process system [D]. Xi'an; Northwestern Polytechnical University, 2012.
2. ZHOU X. Research on dynamic response prediction and control in the milling process of ring-shaped [D]. Xi'an; Northwestern Polytechnical University, 2017.

3. CHOI B K, KIM D H, JERARD R B. C-space approach to tool-path generation for die and mould machining [J]. Computer-Aided Design, 1997, 29(9): 657–669.
4. MORISHIGE K, TAKEUCHI Y, KASE K. Tool path generation using C-space for 5-axis control machining [J]. Journal of Manufacturing Science and Engineering, 1999, 121(1): 144–149.
5. HAN F Y. Research on evoluation modeling method for multi-form intermediate process models and its application [D]. Xi'an; Northwestern Polytechnical University, 2016.
6. GUO H, FU X Y, CHEN F, et al. As-rigid-as-possible shape deformation and interpolation [J]. Journal of Visual Communication and Image Representation, 2008, 19(4): 245–255.
7. HOFFMAN J D, FRANKEL S. Numerical methods for engineers and scientists [M]. 2nd edition. Boca Raton: CRC Press, 2001.
8. GORDON W J, HALL C A. Construction of curvilinear coordinate systems and applications to mesh generation [J]. International Journal for Numerical Methods in Engineering, 1973, 7(4): 461–477.
9. ERIKSSON L. Generation of boundary-conforming grids around wing-body configurations using transfinite interpolation [J]. AIAA Journal, 1982, 20(10): 1313–1320.
10. THOMPSON J F, WARSI Z U A, MASTIN C W. Numerical grid generation, fundamental and application [M]. Amsterdam: North-Holland: Elsevier, 1985.
11. ALAN S, BUDAK E, ÖZGüVEN H N. Analytical prediction of part dynamics for machining stability analysis [J]. International Journal of Automation Technology, 2010, 4(3): 259–267.
12. LIU Y L. Analysis of milling dynamics and chatter detection and control methods for thin-walled workpiece [D]. Xi'an; Northwestern Polytechnical University, 2017.
13. RAO S S. Vibration of continuous systems [M]. New Jersey: John Wiley & Sons, 2007.
14. HOU Y F, ZHANG D H, WU B H, et al. Milling force modeling of worn tool and tool flank wear recognition in end milling [J]. IEEE/ASME Transactions on Mechatronics, 2015, 20(3): 1024–1035.

Chapter 3
Machining Process Monitoring and the Data Processing Method

In the study of metal cutting, the data related to cutting process are generally obtained by the cutting test. For example, the cutting force coefficients used in cutting force prediction are usually obtained through a cutting test. However, the machining process system structure, tool and workpiece material status during the actual machining process are not consistent with the test environment, which means that the cutting parameters obtained under the test conditions are not fully applicable to the real cutting environment. This phenomenon is particularly pronounced during the machining of thin-walled parts. For example, during the machining of blade-type parts, the removed volume of the workpiece material is very large, and the modal parameters of the part vary greatly, and the chatter stability determined according to the initial state of the part alone cannot be applied to the complete cutting process of the workpiece. In the era of intelligent manufacturing, with the development of advanced sensing technology, it is possible to obtain cutting data consistent with the real process system environment and material status in the field through the application of field detection and data processing methods in the cutting process. In the past, the state and quality of the machining process are usually judged manually. Nowadays, it is committed to the use of field detection data for material status detection, tool and workpiece status identification, to provide a more comprehensive process detection means and methods, in order to achieve the correct processing of each product and improve the consistency of batch products.

D. Zhang et al., *Intelligent Machining of Complex Aviation Components*, Research on Intelligent Manufacturing, https://doi.org/10.1007/978-981-16-1586-3_3

3.1 The Detection Method During Cutting Process

Decision-making in intelligent processing requires real-time monitoring of the machining process to obtain useful information. Online monitoring is directed to the process and is linked to specific equipment and instruments. The quantity monitored or detected during machining can be divided into geometry quantity and physical quantity. As for the geometry quantity, it generally includes the measurement of the geometry and surface roughness of the workpiece, the detection of tool wear and the detection of machine tool accuracy. Among them, geometry, surface roughness, tool wear, machine tool accuracy, etc., can be detected by optical means. In addition, the geometry of the workpiece can also be detected using contact-type instruments such as a probe.

The physical quantities commonly used in cutting processes generally include cutting force, torque, vibration, power, pressure, temperature, displacement, acoustic emission, etc., as shown in Table 3.1.

With the continuous development of sensing technology, the technology of integrating small sensors into machine tools, fixtures, spindles and even tools to monitor the machining process has become the development direction in recent years. Compared with the traditional external monitoring method, the advantage of the integrated and embedded monitoring method are:

(1) The monitoring of the complete process can be achieved without interference with the process, which can reflect the practical machining process;
(2) Different machining conditions, different batches and product differences can be monitored to overcome the limitations under the test environment, in which only limited data can be seleceted.

3.2 Machining Process Detection

In the actual milling process, there are many factors affecting the machining process. The influencing factors are mainly divided into four aspects: machining process system, workpiece information, tool status and dynamic parameters. The machining process system includes machine tool performance indicators such as machine tool power, torque, and accuracy. The performance of the machine tool does not change in a short time, that is, the machining system is a time-invariant factor in the machining process. The workpiece information includes workpiece material, part feature, processing method and other indicators related to the workpiece. The workpiece information has been determined before machining and is a time-invariant factor during machining. The tool status includes the geometric parameters, physical parameters and tool wear status information of the tool and the cutting edge. The tool has been determined before machining, and the physical parameters of the tool are time-invariant during the machining process. Tool wear status changes with the progress of the machining process, that is, tool wear status information

Table 3.1 Monitoring physical quantities commonly used in machining process

Monitoring physical quantity	Description
Cutting force	The monitoring of the cutting force is one of the most commonly used monitoring methods for the analysis and study of the cutting process. Depending on the cutting force monitored, the cutting status of the material, the wear state of the tool, etc., can be tracked. The monitoring of cutting forces is generally using force sensors, which can be integrated into the tool holder, fixture or machine tool.
Torque	In the process of drilling, milling, threading, etc., torque is usually required to be monitored, and then according to the monitoring data to determine the machining status.
Vibration	In the cutting process, vibration, tool damage, mechanical collision, etc. will produce abnormal vibration. System vibrations, tool damage, mechanical collisions and serious process failures can be monitored by monitoring vibrations generated by machine structures or workpieces during cutting. Vibration sensors can also monitor the vibration of the machine spindle, providing data for vibration prediction and control during machining.
Power	The power of the machine tool spindle or drive motor during machining can be obtained by the power sensor, and according to the monitored power data, the fluctuations, spikes and short-term drops in power changes can be judged and predicted. Most power monitoring methods are used to prevent spindle overload and to monitor collisions.
Pressure	The pressure detection during machining is mainly to monitor the operating status of the cooling system. Stable coolant pressure is important for the cutting of aeronautical difficult-to-cut materials.
Temperature	Every cutting process produces significant cutting heat, and monitoring of the cutting temperature in machining plays an important role in ensuring the integrity of the machining surface and analyzing the state of the tool. The temperature monitoring method in the cutting process mainly uses the two principles of thermal conduction (thermocouple) and thermal radiation (infrared).
Displacement	The detection of displacement can be achieved by an eddy current displacement sensor or laser displacement sensor, according to displacement, the workpiece vibration status can be analyzed.
Acoustic emission	During cutting, tool breakage produces an acoustic emission signal. The acoustic emission sensor can be used to monitor the acoustic emission signal generated during cutting, and the tool information can be obtained by analyzing the monitored signal. During machining, the sensor can monitor very small acoustic emission signals, which can be combined with effective power or spindle torque to monitor the tool breakage.

is time-varying. Tool wear can also cause changes in the geometric parameters of the cutting edge on the tool. Dynamic parameters include axial cutting depth, radial cutting width, feed rate, and spindle speed. Due to the structure of the workpiece and the deformation of the tool and workpiece during the actual machining, the axial depth of cut and the radial width of cut will change during the machining. At the same time, the feed rate and spindle speed will be optimized and adjusted during the machining process, resulting in changes in the feed speed and spindle speed during the machining process. Therefore, dynamic parameters are time-varying during the machining process.

The time-varying operating condition factors cannot be accurately predicted before the machining process due to its time-varying and complexity. Therefore, it must be acquired through the online identification method during the machining process. However, the influence of time-varying operating conditions on the observables in the machining process is coupled together, and it is difficult to decouple. By monitoring the observables in the processing process, the coupling influence of various time-varying conditions on the observables cannot be decoupled, that is, the traditional methods cannot realize the online identification of time-varying conditions. To this end, a detection processing method is provided for the identification of complex machining conditions in the process [1].

3.2.1 The Concept of Detection Processing

The so-called detection processing is to actively stimulate the machining process in the actual machining process, and at the same time monitor the observable measurement, to obtain the system response before and after the excitation. On this basis, the detection response equation is established according to the process model of the machining process and the system response. After the detection response equation is obtained through detection processing, the response equation is solved to obtain the current time-varying operating conditions of the machining system. The principle of detection process is shown in Fig. 3.1.

"Detection" refers to the purposeful active stimulation of the machining process, and the specific form of active stimulation is determined according to the machining condition factors that need to be identified.

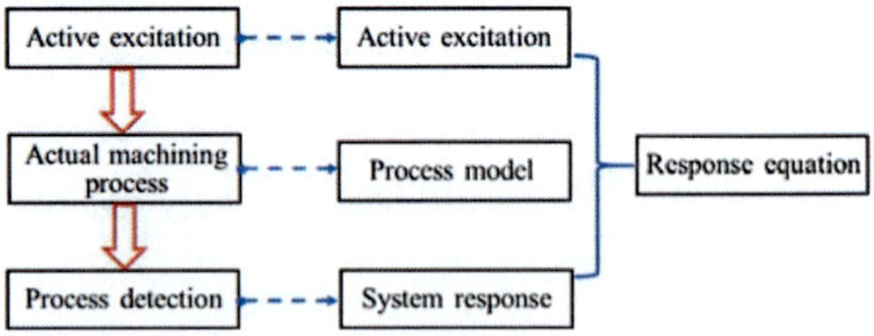

Fig. 3.1 Detection processing principle

"Measurement" in detection processing refers to online monitoring and acquisition of multi-field coupling information on the processing interface of the machining system through sensing means, and the output response of the system under the multi-field coupling action.

"Processing" in detection processing refers to the actual part cutting process, rather than trial cutting or process simulation, it is the carrier of detection processing. In this real processing process, it is necessary to realize the purposeful active excitation according to the specific form. At the same time, it needs to be monitored online to achieve real-time acquisition of the output response of the process system.

The realization method of active excitation in detection processing: During the machining, through communication with the numerical control system, an excitation increment is superimposed on the parameters to realize the active excitation of the processing process. The realization method of process detection during machining is: for the observable measurement in the process, it is measured online through a variety of sensing means, and the measured data is analyzed and processed in real-time to obtain the output response of the process system. Before processing, it is necessary to complete the modeling of the specific machining process. The detection response equation is a specific case of the machining process model at two moments before and after detection. This example is constructed on the basis of the machining process model based on the detection response of the machining process. The detection response equation is solved online, and the current time-varying working condition of the process system can be calculated according to the influence of working condition factors on the observation.

3.2.2 Implementation Method of Detection Processing

3.2.2.1 Mathematical Expression of Detection Processing

In order to realize the detection of the machining process, it is necessary to describe the machining process and establish a machining process model. The generalized expression of the machining process model is:

$$\boldsymbol{M} = f(\boldsymbol{E}) \tag{3.1}$$

where: $\boldsymbol{M} = [m_1\ m_2\ m_3\ ...]^T$ is the observable system output vector during processing; $\boldsymbol{E} = [e_1\ e_2\ e_3...]^T$ is the input condition vector of the machining process; f is the mapping relationship between input conditions and system output, obtained by modeling the machining process.

Excitation increment: Actively superimpose an excitation increment $\Delta\boldsymbol{E}$ on the input condition vector $\boldsymbol{E}$ during machining to form a new input condition vector $\boldsymbol{E} + \Delta\boldsymbol{E}$. The excitation increment vector is $\Delta\boldsymbol{E} = [\Delta e_1\ \Delta e_2\ \Delta e_3...]^T$, and the increment of each element needs to be determined according to the actual detection and recognition process. In the actual detection and recognition process, the elements in the excitation increment vector $\Delta \boldsymbol{E}$ can be partially zero.

System response: The machining process is monitored online by sensing means, and the system output vector $\boldsymbol{M}$ before excitation and the system output vector $\boldsymbol{M}+\Delta \boldsymbol{M}$ after excitation are obtained, respectively. $\Delta \boldsymbol{M} = [\Delta m_1 \ \Delta m_2 \ \Delta m_3 \ \cdots]^T$ is the excitation response vector of the system. This vector can be comparing the system output after excitation before excitation.

Detection process: In the real machining process, actively superimpose an excitation increment $\Delta \boldsymbol{E}$ to the input condition vector to form a new input condition vector $\boldsymbol{E} + \Delta \boldsymbol{E}$, and monitor the output vector $\boldsymbol{M}$ of the machining system before excitation and the output vector of the process system after excitation $\boldsymbol{M} + \Delta \boldsymbol{M}$, get the output response of the process system. According to the established machining process model, that is, the mapping relationship $f : \boldsymbol{E} \rightarrow \boldsymbol{M}$ from input conditions to system output, the detection response equation of the system is constructed.

Detection response equation: Based on the process model, the process input condition vector $\boldsymbol{E}$, system output vector $\boldsymbol{M}$ and the process input condition vector $\boldsymbol{E}+\Delta \boldsymbol{E}$ and system output vector $\boldsymbol{M} + \Delta \boldsymbol{M}$ before active excitation are respectively brought into the process model. The resulting detection response equation. Its expression is:

$$\begin{cases} \boldsymbol{M}_i = f(\boldsymbol{E}_i) \\ \boldsymbol{M}_i + \Delta \boldsymbol{M}_i = f(\boldsymbol{E}_i + \Delta \boldsymbol{E}_i) \end{cases} \tag{3.2}$$

Online recognition of time-varying machining conditions: According to the specific requirements of monitoring and identifying time-varying machining conditions, the Eq. (3.2) is solved online to obtain the model coefficients in the process model $f : \boldsymbol{E} \rightarrow \boldsymbol{M}$. According to the influence of working condition factors on model coefficients, the current time-varying condition factors of the process system can be calculated, so as to realize the online monitoring of time-varying machining condition factors.

From the perspective of the machining process, the detection of the machining process has been compared with the output of the system before and after the active excitation, that is, the detection response of the system. On this basis, further online analysis of the machining process can be carried out. From the point of view of mathematical models, the detection of the machining process increases the number of equations, turning the unsolvable indeterminate equations into solvable exact equations or overdetermined equations. On this basis, the time-varying machining condition factors can be obtained by solving the equations.

3.2.2.2 The Realization of Detection Processing

The detection process can be divided into three stages: pre-preparation, online detection and post-processing. The main process is shown in Fig. 3.2.

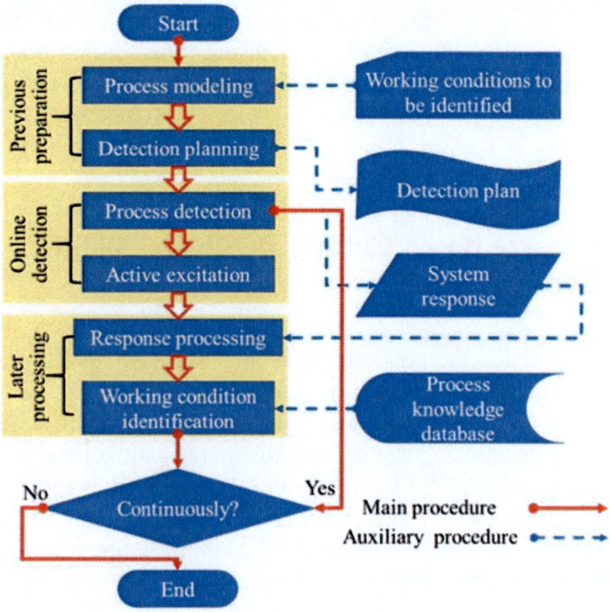

Fig. 3.2 Detection process

(1) Process modeling. Analyze the machining process, and establish the relevant machining process model according to the working condition factors that need to be identified, which provides a theoretical basis for the construction and solution of the detection process and the response equation.
(2) Detection planning. Develop a detection plan based on the real part machining process and the machining condition factors that need to be identified, including the main excitation form, process detection method, and hardware implementation of the detection process.
(3) Process detection. In the real part machining process, in accordance with the process monitoring plan that has been formulated, through a variety of sensing means, the observable measurement in the process is monitored in real time, and the system response of the machining process is obtained.
(4) Active excitation. In the real part machining process, according to the active excitation form that has been established, the machining process is controlled through active communication with the numerical control system, and the process of excitation is increased.
(5) Response processing. After the detection process is completed, the detection response results obtained from the monitoring are analyzed and processed online, the detection response equation is constructed according to the machining process model, and the response equation is initially solved.
(6) Machining condition identification. Obtain the influence relationship of the time-varying machining condition factors on the observable measurement

of the machining process through the matching of the process knowledge base, and calculate and identify the current time-varying machining condition according to the influence relationship and the preliminary solution result of the detection response equation.

3.3 Milling Force Based Cutting Depth and Width Detection

In the detection process, the nominal depth and width of cut are generally determined when planning the toolpath. However, the inconsistency of parts in the actual machining process (especially the rough machining stage), workpiece deformation and other factors will cause the actual depth of cut, the width of cut and other parameters to be inconsistent with the nominal values, which will affect subsequent parts processing and optimization. Therefore, it is necessary to identify the actual milling parameters based on the online monitoring of the milling force and other data, thus to lay the foundation for the optimization of the complex part machining.

In the milling process, factors such as the physical properties of the workpiece material, the parameters and wear conditions of the tool, and the milling parameters (feed per tooth, axial depth of cut, radial width of cut) and other factors will simultaneously affect the milling force. In the milling process, the wear state of the tool and the milling parameters change with the progress of milling. Therefore, the time-varying machining conditions that affect the milling force include the wear state of the tool and the milling parameters. The milling parameters and the wear status of the tool simultaneously affect the milling force, that is, the milling force cannot reflect the milling parameters or the wear status of the tool alone. Therefore, it is necessary to decouple the influence of multiple factors through detection methods, and then perform online identification of time-varying machining conditions.

3.3.1 Average Milling Force

In the milling process, the physical quantity most related to the depth and width of cut is the milling force, and the milling force signal has the characteristics of fast response, high sensitivity, and can be collected online. Therefore, in the process of identifying the depth of cut and width of cut, it is appropriate to use milling force as the object to be monitored in the detection process.

According to the milling force model considering tool wear and analysis of worn tools on the milling force in the case of changing machining conditions [1], the average milling force can be expressed as a linear function of the feed per tooth f_z:

$$\overline{\boldsymbol{F}} = f_z \cdot \boldsymbol{G}_c \cdot \boldsymbol{K} + \boldsymbol{G}_w \cdot \boldsymbol{F}_w(VB) \tag{3.3}$$

where $\overline{\boldsymbol{F}}$ is the average milling force vector, $\boldsymbol{G}_c$ is the geometric influence matrix of the average shear force in the average milling force, $\boldsymbol{K}$ is the shear force coefficient matrix, f_z is the feed per tooth, and $\boldsymbol{G}_w$ is the geometric influence matrix of the average friction effect in the average milling force, $\boldsymbol{F}_w(VB)$ is the tool wear influence matrix of the milling force. In the recognition process of the depth of cut and width of cut described in this section, the impact of tool wear is temporarily ignored.

3.3.2 Detection and Measurement in the Milling Process

In the real milling process, the milling force is generally monitored by a cutting force sensor, and the average milling force is calculated in real-time. The calculated average milling force is denoted as $\overline{\boldsymbol{F}_{moni}}$. When it is necessary to identify the current axial depth of cut a_p and radial width of cut a_e, the numerical control system actively superimposes an excitation increment Δf_z to the feed per tooth f_z, so that the actual feed per tooth is $f_z + \Delta f_z$. At the same time, it monitors the real-time milling force and calculates the average milling force $\overline{\boldsymbol{F}_{moni}} + \Delta\overline{\boldsymbol{F}_{moni}}$. In this milling detection process, "detection" refers to actively superimposing an excitation increment Δf_z on the feed per tooth f_z to make the actual feed per tooth. The quantity is $f_z + \Delta f_z$. The "measurement" means to monitor the milling force before and after the active excitation through the force sensors, and calculate the corresponding average milling forces $\overline{\boldsymbol{F}_{moni}}$ and $\overline{\boldsymbol{F}_{moni}} + \Delta\overline{\boldsymbol{F}_{moni}}$ in real time. Among them, the average milling force $\overline{\boldsymbol{F}_{moni}}$ before the active excitation and the average milling after the active excitation The force change $\Delta\overline{\boldsymbol{F}_{moni}}$ can be expressed as:

$$\begin{cases} \overline{\boldsymbol{F}_{moni}} = \begin{bmatrix} \overline{\boldsymbol{F}_{moni,x}} \\ \overline{\boldsymbol{F}_{moni,y}} \end{bmatrix} \\ \Delta\overline{\boldsymbol{F}_{moni}} = \begin{bmatrix} \Delta\overline{\boldsymbol{F}_{moni,x}} \\ \Delta\overline{\boldsymbol{F}_{moni,y}} \end{bmatrix} \end{cases} \tag{3.4}$$

3.3.3 Detection Response Equation

According to the average milling force expression in the form of linear function of the feed per tooth f_z, the following two equations can be obtained:

$$\begin{cases} \overline{\boldsymbol{F}_{moni}} = f_z \cdot \boldsymbol{G}_c \cdot K + \boldsymbol{G}_w \cdot \boldsymbol{F}_w(VB) \\ \overline{\boldsymbol{F}_{moni}} + \Delta\overline{\boldsymbol{F}_{moni}} = (f_z + \Delta f_z) \cdot \boldsymbol{G}_c \cdot K + \boldsymbol{G}_w \cdot \boldsymbol{F}_w(VB) \end{cases} \tag{3.5}$$

In this equation group, the average milling force $\overline{\boldsymbol{F}_{moni}}$ before active excitation and the average milling force change $\Delta\overline{\boldsymbol{F}_{moni}}$ after active excitation are monitored and calculated in real time by the load cell. The feed per tooth f_z before active monitoring can be obtained in real time by communicating with the numerical control system. The excitation increment Δf_z is to increase the excitation quantity actively. The shear force coefficient matrix $\boldsymbol{K}$ is calibrated in advance through cutting experiments and stored in the process knowledge base. The geometric influence matrix $\boldsymbol{G}_c$ of the average shear force and the geometric influence matrix $\boldsymbol{G}_w$ of the average friction effect force are both related to the tool parameters, the engagement state of milling, and the depth and width of cut. The tool wear influence matrix $\boldsymbol{F}_c(VB)$ of milling force is related to factors such as tool wear status, workpiece material and tool coating.

3.3.4 Detection and Recognition of Depth and Width of Cut

3.3.4.1 Identification Process Analysis

In Eq. (3.5), unknowns include the geometric influence matrix $\boldsymbol{G}_c$ of the average shear force, the geometric influence matrix $\boldsymbol{G}_w$ of the average friction force, and the tool wear effect matrix $\boldsymbol{F}_w(VB)$ of the milling force. Matrix $\boldsymbol{G}_c$ and matrix $\boldsymbol{G}_w$ are related to tool parameters, tool engagement state, cutting depth and width and other factors, so the two matrices are related. The two matrices are related by their correlation with the cutting depth and width. At the same time, the geometric influence matrix $\boldsymbol{G}_w$ of the average friction effect force in the equation system and the tool wear influence matrix $\boldsymbol{F}_w(VB)$ of the milling force are cross-coupled together, which is difficult to solve separately. Therefore, the cutting depth and width cannot be identified by the geometric influence matrix $\boldsymbol{G}_w$ calculation of the average frictional force, but only by the geometric influence matrix $\boldsymbol{G}_c$ calculation of the average shear force.

According to Eq. (3.5), the item $\boldsymbol{G}_c \cdot \boldsymbol{K}$ of the geometric influence matrix $\boldsymbol{G}_c$, which contains the average shear force, can be expressed as:

$$\boldsymbol{G}_c \cdot \boldsymbol{K} = \Delta \overline{\boldsymbol{F}_{moni}} / \Delta f_z \quad (3.6)$$

The geometric influence matrix $\boldsymbol{G}_c$ of the average shear force can be expressed as:

$$\boldsymbol{G}_c = \begin{bmatrix} g_{c,1} & g_{c,2} \\ -g_{c,2} & g_{c,1} \end{bmatrix} \quad (3.7)$$

where: $g_{c,1}$ and $g_{c,2}$ can be expressed as:

$$\begin{cases} g_{c,1} = \frac{N \cdot a_p}{8\pi} \cdot (cos2\varphi_{j,st} - cos2\varphi_{j,ex}) \\ g_{c,2} = \frac{N \cdot a_p}{8\pi} \cdot (sin2\varphi_{j,st} - sin2\varphi_{j,ex} - 2(\varphi_{j,st} - \varphi_{j,ex})) \end{cases} \quad (3.8)$$

among them, N is the number of mill teeth, $\varphi_{j,st}$ is the entry angle of the milling cutter teeth, and $\varphi_{j,ex}$ is the exit angle of the milling cutter teeth.

The shear force coefficient matrix $\boldsymbol{K}$ can be expressed as:

$$\boldsymbol{K} = \begin{bmatrix} K_{tc} \\ K_{rc} \end{bmatrix} \tag{3.9}$$

Bringing matrix $\boldsymbol{G}_c$, matrix $\boldsymbol{K}$, and matrix $\Delta\overline{\boldsymbol{F}_{moni}}$ into the Eq. (3.6) at the same time, $g_{c,1}$ and $g_{c,2}$ can be solved respectively, and the result is:

$$\begin{bmatrix} g_{c,1} \\ g_{c,2} \end{bmatrix} = \frac{1}{\Delta f_z \cdot \left(K_{tc}^2 + K_{rc}^2\right)} \cdot \begin{bmatrix} \Delta\overline{\boldsymbol{F}_{moni,x}} & \Delta\overline{\boldsymbol{F}_{moni,y}} \\ -\Delta\overline{\boldsymbol{F}_{moni,y}} & \Delta\overline{\boldsymbol{F}_{moni,x}} \end{bmatrix} \cdot \boldsymbol{K} \tag{3.10}$$

where Δf_z is the active excitation in the detection process. It is determined by the detection process, and it is a known quantity. $\boldsymbol{K}$ is the shear force coefficient matrix, including K_{tc}, K_{rc} two components, the shear force coefficient matrix $\boldsymbol{K}$ can be matched in the process knowledge base.Once the workpiece and tool materials are determined, $\Delta\overline{\boldsymbol{F}_{moni}, x}$ and $\Delta\overline{\boldsymbol{F}_{moni}, y}$ are the average milling force change after active excitation in the X direction and Y direction, they can be monitored and calculated in real time by the force sensor. After calculating $g_{c,1}$ and $g_{c,2}$, the geometric influence matrix $\boldsymbol{G}_c$ of the average shear force can be obtained.

3.3.4.2 The Computational Identification for Cutting Depth and Width

The geometric influence matrix $\boldsymbol{G}_c$ of the average shear force is related to tool parameters, cutting depth and width, and other factors. After the solution response equation obtains the independent elements $g_{c,1}$ and $g_{c,2}$, the actual a_p and a_e in the milling process can be calculated according to the effect of the axial cutting depth a_p and radial cutting width a_e on $g_{c,1}$ and $g_{c,2}$. In Eq. (3.8), the number of milling cutter teeth is fixed. The axial milling depth and radial cutting width change in real time during milling and cannot be accurately predicted prior to machining.

Take the up-milling process as an example, according to Eq. (3.8), the following can be derived:

$$\frac{g_{c,2}}{g_{c,1}} = \frac{sin2\varphi_{j,st} - sin2\varphi_{j,ex} - 2\left(\varphi_{j,st} - \varphi_{j,ex}\right)}{cos2\varphi_{j,st} - cos2\varphi_{j,ex}} \tag{3.11}$$

In up-milling process $\varphi_{j,st} = 0$, Eq. (3.11) can be reduced to:

$$\frac{g_{c,2}}{g_{c,1}} = \frac{-sin\varphi_{j,ex} \cdot cos\varphi_{j,ex} + \varphi_{j,ex}}{sin^2\varphi_{j,ex}} \tag{3.12}$$

Define the function $f(x)$:

$$f(x) = \frac{-sin\, x \cdot cos\, x + x}{sin^2 x} \tag{3.13}$$

Function $f_{(x)}$ monotonically increments on intervals $(0, \pi)$. The minimum value of function $f_{(x)}$ on interval $(0, \pi)$ is $\lim_{x\to 0} f_{(x)} = 0$, and the maximum value of function $f_{(x)}$ is $\lim_{x\to \pi} f_{(x)} = +\infty$. According to the above conclusion, the right side of the Eq. (3.12) medium is a monotonous increment function of $\varphi_{j,ex}$. At this point, the unique $\varphi_{j,ex}$ can be determined according to the expression of in the right side of the Eq. (3.12), and the only a_p corresponding to it can be obtained by submitting $\varphi_{j,ex}$ in to Eq. (3.8). According to the relationship between angle and radial cutting width $\varphi_{j,ex} = arcos[(R - a_e)/R]$, the only a_e corresponding to it can be setted.

In summary, it can be obtained that when a set of $g_{c,1}$ and $g_{c,2}$ is given, there is and the only set of a_p, a_e corresponds to it, that is, the mapping relationship between $g_{c,1}$ and $g_{c,2}$ and axial milling depth a_p and radial cutting width a_e can be established. Therefore, the a_p and a_e in the milling process can be obtained through the mapping relationships of $g_{c,1}$, $g_{c,2}$ to a_p, a_e. Based on the mapping relationships from $g_{c,1}$ and $g_{c,2}$ to a_p and a_e in the discrete from, the interpolation results are shown in Fig. 3.3.

At this point, during the machining process, by the active excitation of the feed per tooth f_z, while monitoring the milling force before and after the active excitation, the detected response equations can be obtained. According to the response equation, the geometric influence matrix $\boldsymbol{G}_c$ of the average shear force can be calculated, and the inverse mapping of the influence relationship between a_p and a_e on matrix $\boldsymbol{G}_c$ can be calculated to get the cutting depth and width in milling process, and the detection and recognition of dynamic depth and width in the milling process can be realized.

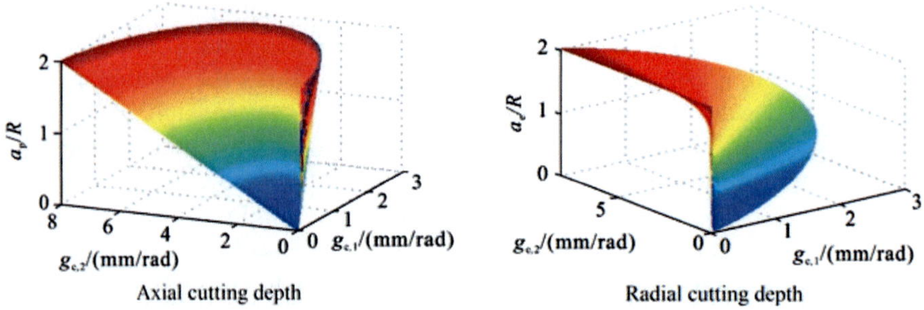

Fig. 3.3 $g_{c,1}$, $g_{c,2}$ to the deep-wide mapping relationship

3.4 Detection and Recognition of Milling Cutter Wear Status

In the milling process, the time-varying machining condition factors that affect the milling force include the wear state of the tool and the milling parameters. Among them, the milling parameters (feed per tooth, axial depth of cut, radial cutting width) can be set during programming or online recognition, that is, the coupling influence of milling parameters and tool wear status on milling force can be decoupled. According to the milling force model, the milling parameters can be further separated from the coupling terms of the milling parameters and the tool wear status, and the effect of the tool wear status on the milling force can be solved according to the identified milling parameters. By calculating the influence items of the tool wear status on the milling force, the tool wear status can be obtained.

3.4.1 Measurement of Tool Wear

In industrial production, the measurement of tool wear status mainly include direct measurement and indirect measurement. The direct measurement method generally uses some measurement methods to directly measure the wear area of the tool. The most commonly used method is the optical measurement method. Measurement methods based on optical principles generally have very good measurement accuracy, but the measurement process and data processing workload are relatively large, and the measurement efficiency is low. In addition, when measuring tool wear in the milling process, this method is susceptible to interference from cutting fluid, chips, etc., resulting in inaccurate measurement results. The indirect measurement method generally indirectly infers the wear state of the tool by monitoring some cutting process signals related to the tool wear process, such as measuring the cutting force, vibration, power and other signals in the machining process related to tool wear. The advantage of the indirect measurement method is that it can realize the rapid monitoring of the online status, but due to the influence of other non-wear factors, its accuracy is not as high as that of the direct measurement method. From the point of view of industrial field application, the indirect measurement method is more suitable for monitoring tool wear during processing. This is also the focus of the development of academic and industrial in recent years.

3.4.2 Milling Force Model of Worn Tool

3.4.2.1 Milling Force Model with Tool Wear

In the milling process, the shearing force caused by the shearing action on the rake face and the friction and extrusion force caused by the wear of the flank face together acts on the tool to form a milling force. Here, the friction and squeezing force caused by the flank wear are collectively referred to as the friction effect force. According to the existing milling force model, the shearing force is related to the undeformed cutting thickness of the material, and the influence of flank wear on it is not considered. The friction effect force is produced by the friction and extrusion between the flank surface and the processed surface. Therefore, the friction effect force is related to the wear of the tool flank surface and has nothing to do with the undeformed chip thickness of the material. The milling force analysis of the cutter tooth element is shown in Fig. 3.4.

1. **Shear force model**

Discrete the cutter into cutter tooth micro-elements along the cutter axis. The shearing force on the micro-element of the *j*-th tooth of the milling cutter with the cutter rotation angle φ and the height z can be expressed as [2]:

$$\begin{cases} dF_{j,tc}\left(\varphi_j(z)\right) = K_{tc}h\left(\varphi_j(z)\right)dz \\ dF_{j,rc}\left(\varphi_j(z)\right) = K_{rc}h\left(\varphi_j(z)\right)dz \\ dF_{j,ac}\left(\varphi_j(z)\right) = K_{ac}h\left(\varphi_j(z)\right)dz \end{cases} \tag{3.14}$$

where $dF_{j,tc}\left(\varphi_j(z)\right)$, $dF_{j,rc}\left(\varphi_j(z)\right)$, $dF_{j,ac}\left(\varphi_j(z)\right)$ are the tangential, radial and axial shear forces experienced by the tool respectively, K_{tc}, K_{rc}, and K_{ac} are the tangential, radial and axial shear force coefficients respectively, $h\left(\varphi_j(z)\right)$ is the undeformed chip thickness of the tooth at the angle $\varphi_j(z)$, $\varphi_j(z)$ is the rotation angle of the *j*th tooth of the axial micro-element at the height z. The infinitesimal shear force can be transformed into the $OXYZ$ coordinate system.

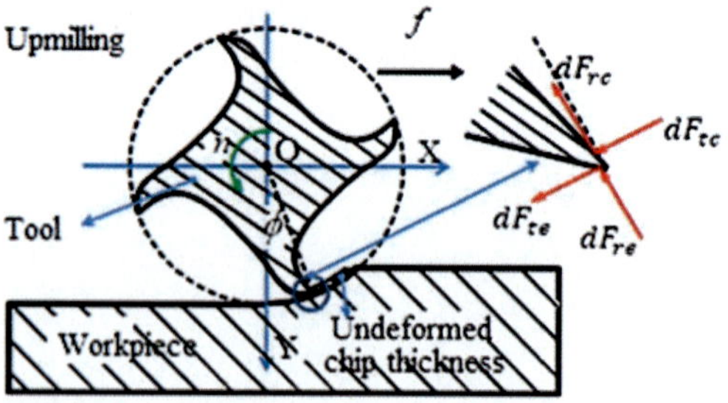

Fig. 3.4 Analysis of micro-milling force of tooth

2 Friction effect force model

According to Teitenberg's theory [3], the effect of flank wear on the axial milling force can be ignored, so only the force in the OXY plane is considered in the friction effect force model. The friction and squeezing force on the tool micro-element can be expressed as a function of the tool flank wear VB, namely:

$$\begin{cases} dF_{tw} = F_{tw}(VB)dz \\ dF_{rw} = F_{rw}(VB)dz \end{cases} \tag{3.15}$$

where: dF_{tw} and dF_{rw} are the tangential friction force and the normal squeezing force of a single cutter tooth on the micro-element dz, respectively; $F_{tw}(VB)$ and $F_{rw}(VB)$ are the friction force and squeezing force on the unit edge length respectively, both of which are related to the tool flank wear VB.

3.4.2.2 The Distribution of Frictional Force

The friction and squeezing force per unit edge length can be expressed as:

$$\begin{cases} F_{tw}(VB) = \int_0^{VB} \tau(x)dx \\ F_{rw}(VB) = \int_0^{VB} \sigma(x)dx \end{cases} \tag{3.16}$$

where $\tau(x)$ and $\sigma(x)$ are the shear stress and normal stress at the distance x between the flank surface and the cutting edge, respectively.

As shown in Fig. 3.5, according to the research of Lapsley [4] and Waldorf [5], the contact area between the worn flank and the workpiece material is divided into a plastic flow area and an elastic contact area. In the plastic flow zone, the shear stress

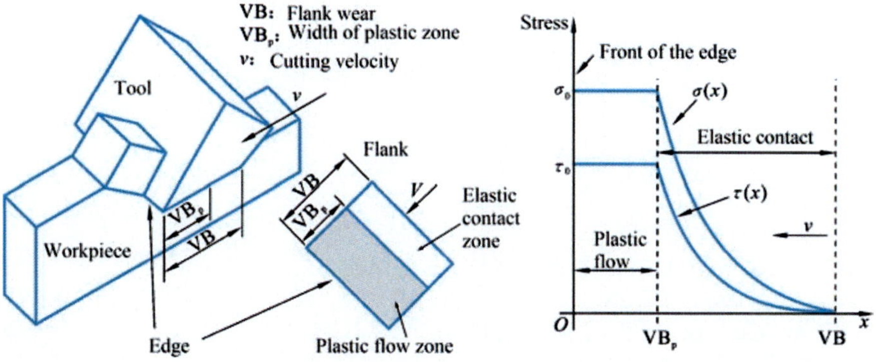

Fig. 3.5 Flank wear and stress distribution

and the normal stress are both constant values τ_0 and σ_0. In the elastic contact zone, the stress is distributed according to the quadratic law [6].

When $0 < x < VB_p$, the contact area is in the plastic flow zone, and its stress distribution is:

$$\begin{cases} \tau(x) = \tau_0 \\ \sigma(x) = \sigma_0 \end{cases} \tag{3.17}$$

When $VB_p < x < VB$, the contact area is in the elastic contact area, and its stress distribution is:

$$\begin{cases} \tau(x) = \tau_0\left(\frac{VB-x}{VB-VB_P}\right)^2 \\ \sigma(x) = \sigma_0\left(\frac{VB-x}{VB-VB_P}\right)^2 \end{cases} \tag{3.18}$$

where x is the distance from a certain point on the flank surface to the cutting edge.

According to Smithey's research [7], when the wear reaches a certain amount, the width of the elastic contact zone remains unchanged, and the width of the plastic flow zone increases with the increase in the amount of wear, namely:

$$VB_P = \begin{cases} 0 & VB < VB^* \\ VB - VB^* & VB \geq VB^* \end{cases} \tag{3.19}$$

where VB^* is the fixed width of the elastic contact area.

Substitute Eq. (3.17) to Eq. (3.19) into Eq. (3.16), and the friction force and squeezing force per unit edge length can be obtained by integration.

When $VB < VB^*$, there is:

$$\begin{cases} F_{tw}(VB) = \frac{\tau_0}{3} \cdot VB \\ F_{rw}(VB) = \frac{\sigma_0}{3} \cdot VB \end{cases} \tag{3.20}$$

When $VB \geq VB^*$, there is:

$$\begin{cases} F_{tw}(VB) = \tau_0\left(VB - \frac{2}{3}VB^*\right) \\ F_{rw}(VB) = \sigma_0\left(VB - \frac{2}{3}VB^*\right) \end{cases} \tag{3.21}$$

3.4.2.3 Average Milling Force of Worn Cutter

1. **The resultant force of worn milling cutter**

The resultant force acting on the tool as a whole is the sum of the shearing force and the frictional force, namely:

$$F = F_c + F_w \tag{3.22}$$

By integrating the force on each cutter tooth element within the range of the cutter's axial depth of cut $\left[z_{j,1}(\varphi_j), z_{j,2}(\varphi_j)\right]$, and then the shear force and friction effect force of each cutter tooth can be obtained:

$$\begin{cases} F_{j,xc}(\varphi_j) = \frac{f_z}{4k_\beta}\{-K_{tc}\cos(2\varphi_j(z)) + K_{rc}(2\varphi_j(z) - \sin(2\varphi_j(z)))\}_{z_{j,1}(\varphi_j)}^{z_{j,2}(\varphi_j)} \\ F_{j,yc}(\varphi_j) = -\frac{f_z}{4k_\beta}\{K_{tc}(2\varphi_j(z) - \sin(2\varphi_j(z))) + K_{rc}\cos(2\varphi_j(z))\}_{z_{j,1}(\varphi_j)}^{z_{j,2}(\varphi_j)} \\ F_{j,zc}(\varphi_j) = \frac{f_z}{k_\beta}\{K_{ac}\cos(\varphi_j(z))\}_{z_{j,1}(\varphi_j)}^{z_{j,2}(\varphi_j)} \end{cases} \tag{3.23}$$

$$\begin{cases} F_{j,xw}(\varphi_j) = \frac{1}{k_\beta}\{F_{tw}(VB)\sin(\varphi_j(z)) - F_{rw}(VB)\cos(\varphi_j(z))\}_{z_{j,1}(\varphi_j)}^{z_{j,2}(\varphi_j)} \\ F_{j,yw}(\varphi_j) = \frac{1}{k_\beta}\{F_{tw}(VB)\cos(\varphi_j(z)) + F_{rw}(VB)\sin(\varphi_j(z))\}_{z_{j,1}(\varphi_j)}^{z_{j,2}(\varphi_j)} \end{cases} \tag{3.24}$$

where $k_\beta = tan\beta/R$ is the geometric coefficient of the cutter, β is the helix angle of the milling cutter, and R is the radius of the milling cutter.

Thus, the resultant force experienced by the milling cutter is:

$$\begin{cases} F_{j,x}(\varphi_j) = F_{j,xc}(\varphi_j) + F_{j,xw}(\varphi_j) \\ F_{j,y}(\varphi_j) = F_{j,yc}(\varphi_j) + F_{j,yw}(\varphi_j) \\ F_{j,z}(\varphi_j) = F_{j,zc}(\varphi_j) \end{cases} \tag{3.25}$$

2. **Average milling force of worn cutter**

Since the flank wear has a small effect on the axial milling force [1], the milling forces in the X and Y directions are averaged, and the expression is:

$$\begin{cases} \overline{F_{j,x}} = \overline{F_{j,xc}} + \overline{F_{j,xw}} \\ \overline{F_{j,y}} = \overline{F_{j,yc}} + \overline{F_{j,yw}} \end{cases} \tag{3.26}$$

The expression of the average shear force is:

$$\begin{cases} \overline{F_{j,xc}} = \frac{N}{2\pi} \cdot \int_{\varphi_{j,st}}^{\varphi_{j,ex}+a_p k_\beta} F_{j,xc}(\varphi_j)d\varphi_j \\ \overline{F_{j,yc}} = \frac{N}{2\pi} \cdot \int_{\varphi_{j,st}}^{\varphi_{j,ex}+a_p k_\beta} F_{j,yc}(\varphi_j)d\varphi_j \end{cases} \tag{3.27}$$

The expression of the average friction effect force is:

$$\begin{cases} \overline{F_{j,xw}} = \frac{N}{2\pi} \cdot \int_{\varphi_{j,st}}^{\varphi_{j,ex}+a_p k_\beta} F_{j,xw}(\varphi_j)d\varphi_j \\ \overline{F_{j,yw}} = \frac{N}{2\pi} \cdot \int_{\varphi_{j,st}}^{\varphi_{j,ex}+a_p k_\beta} F_{j,yw}(\varphi_j)d\varphi_j \end{cases} \tag{3.28}$$

where N is the cutter teeth number.

Substituting the calculation expression of the milling force in the milling force model of the worn tool into Eqs. (3.27) and (3.28) to obtain the calculation result. The average milling force of the entire milling cutter is equal to the average milling force of a single tooth. The average milling force of the entire milling cutter can be written as a linear function of the feed per tooth f_z, which is Eq. (3.5).

The geometric influence matrix of the average friction effect force is affected by tool parameters, tool engagement form, machining parameters, etc., and its expression is:

$$\boldsymbol{G}_w = \begin{bmatrix} -g_{w,1} & g_{w,2} \\ -g_{w,2} & -g_{w,1} \end{bmatrix} = \frac{Na_p}{2\pi}\begin{bmatrix} \sin\varphi_{j,ex} - \sin\varphi_{j,st} & \cos\varphi_{j,st} - \cos\varphi_{j,ex} \\ \cos\varphi_{j,ex} - \cos\varphi_{j,st} & \sin\varphi_{j,ex} - \sin\varphi_{j,st} \end{bmatrix} \tag{3.29}$$

The tool wear impact matrix of milling force is affected by tool wear status, tool engagement form, machining parameters, etc., which can be expressed as:

$$\boldsymbol{F}_w(VB) = \begin{bmatrix} F_{tw}(VB) \\ F_{rw}(VB) \end{bmatrix} \tag{3.30}$$

3.4.3 *Identification Process Analysis*

According to Eq. (3.5), the geometric influence matrix $\boldsymbol{G}_w$ containing the average friction effect force in the average milling force and the item $\boldsymbol{G}_w \cdot \boldsymbol{F}_w(VB)$ of the tool wear influence matrix $\boldsymbol{F}_w(VB)$ of the milling force are solved to obtain:

$$\boldsymbol{G}_w \cdot \boldsymbol{F}_w(VB) = \overline{\boldsymbol{F}_{moni}} - f_z \cdot \boldsymbol{G}_c \cdot \boldsymbol{K} \tag{3.31}$$

According to Eq. (3.29), the two elements $g_{w,1}$ and $g_{w,2}$ in the geometric influence matrix $\boldsymbol{G}_w$ of the average friction effect force can be expressed as:

$$\begin{cases} g_{w,1} = \frac{N \cdot a_p}{2\pi} \cdot \left(sin\varphi_{j,st} - sin\varphi_{j,ex}\right) \\ g_{w,2} = \frac{N \cdot a_p}{2\pi} \cdot \left(cos\varphi_{j,st} - cos\varphi_{j,ex}\right) \end{cases} \tag{3.32}$$

According to the definition of matrix $\boldsymbol{G}_w$, matrix $\boldsymbol{G}_w$ is an invertible matrix. Multiplying the inverse matrix $\boldsymbol{G}_w^{-1}$ of the matrix $\boldsymbol{G}_w$ on the left and right sides of the Eq. (3.31) at the same time, yields:

$$\boldsymbol{F}_w(VB) = \boldsymbol{G}_w^{-1} \cdot \left(\overline{\boldsymbol{F}_{moni}} - f_z \cdot \boldsymbol{G}_c \cdot \boldsymbol{K}\right) \tag{3.33}$$

It can be further derived as:

$$\boldsymbol{F}_w(VB) = \frac{1}{g_{w,1}^2 + g_{w,2}^2} \cdot \boldsymbol{G}_w^T \cdot \left(\boldsymbol{F}_{moni} - f_z \cdot \boldsymbol{G}_c \cdot \boldsymbol{K}\right) \tag{3.34}$$

3.4.4 Calculation and Identification of Wear

According to the modeling process of the milling force of the worn tool, the two components of the matrix $\boldsymbol{F}_w(VB)$, namely the friction force and the squeezing force on the unit edge length, $F_{tw}(VB)$ and $F_{rw}(VB)$, have a linear relationship with the flank wear of the milling cutter VB. According to the influence of tool wear on the friction force and the squeeze force $F_{tw}(VB)$, $F_{rw}(VB)$ on the unit cutting edge, the mapping from VB $F_{tw}(VB)$ and $F_{rw}(VB)$ per unit edge length to the milling cutter flank wear can be obtained. VB In the case that the friction force and the pressing force $F_{tw}(VB)$ and $F_{rw}(VB)$ per unit edge length are determined, a flank wear amount VB can be obtained according to $F_{tw}(VB)$ and $F_{rw}(VB)$, respectively. From the perspective of the milling force model, the two flank wears VB obtained should be the same. However, due to a certain error in the detection process, the two flank surface wear amounts VB obtained may be different. That is, $F_{tw}(VB)$ and $F_{rw}(VB)$ detected during the machining process are contradictory to the flank wear VB to $F_{tw}(VB)$ and $F_{rw}(VB)$. Therefore, the least square method is used to solve the contradiction equations composed of the influence relationship of VB on $F_{tw}(VB)$ and $F_{rw}(VB)$.

Firstly, solve Eq. (3.20) by the least square method. Then the range of friction and squeezing force per unit edge length is:

$$\begin{cases} F_{tw}(VB) < \frac{\tau_0}{3} \cdot VB^* \\ F_{rw}(VB) < \frac{\sigma_0}{3} \cdot VB^* \end{cases} \tag{3.35}$$

The result of the least squares solution is:

$$VB = \frac{3[\tau_0 \; \sigma_0] \cdot \boldsymbol{F}_w(VB)}{[\tau_0 \; \sigma_0] \cdot [\tau_0 \; \sigma_0]^T} \tag{3.36}$$

Solve Eq. (3.21) by least squares. At this time, the range of friction and extrusion force per unit blade length is:

$$\begin{cases} F_{tw}(VB) \geq \frac{\tau_0}{3} \cdot VB^* \\ F_{rw}(VB) \geq \frac{\sigma_0}{3} \cdot VB^* \end{cases} \tag{3.37}$$

The result of the least squares solution is:

$$VB = \frac{2}{3} VB^* + \frac{[\tau_0 \; \sigma_0] \cdot \boldsymbol{F}_w(VB)}{[\tau_0 \; \sigma_0] \cdot [\tau_0 \; \sigma_0]^T} \tag{3.38}$$

For the specific combination of tool and workpiece material, the parameters VB^*, τ_0, σ_0 can be obtained by matching in the process knowledge base or obtained through experiments.

3.5 Identification of Cutting Force Coefficients Based on Monitored Data

After the cutting force prediction model has been developed for a long time, the prediction accuracy has been greatly improved, and it is closer to real situations. However, the basis of the existing cutting force model is still the cutting force coefficients describing material deformation characteristics. Therefore, obtaining accurate cutting force coefficients is the premise for establishing a precise prediction model. Many scholars have studied the problems existing in the identification of cutting force coefficients. In addition to the widely used slot milling, orthogonal and oblique cutting experimental methods, many new identification methods of cutting force coefficients have also been developed. The proposed new cutting force coefficient identification methods make the identification of cutting force coefficients operate under complex situations can be realized. In other words, the cutting force coefficients identification in the experimental design can be closer to the actual machining conditions. However, the real machining conditions are extremely complicated, and there are many influencing factors that are as obvious as quantitative elements but cannot be directly quantitatively described. For example, cooling conditions are affected by the openness of actual machining equipment and parts, and the parts are different from specimens in the calibration experiments. In addition, the calibration experiments have some deficiencies, like large workloads, scattered collected data, and repeated tests. Although the data monitored from the actual process has the problems of poor quality, the total amount of data is the most abundant, it has the advantages of wide data sources, and contains rich processing information, which can be used as an important resource for coefficient identification. Therefore, in order to utilize sufficient real-time machining data and consider quantifiable influencing factors and eliminate the influence of non-quantifiable factors, an algorithm that can be applied to online identification and can output a large number of cutting force coefficients for statistical processing is required [8].

3.5.1 Cutting Force Modeling Considering Cutter Vibrations

3.5.1.1 Ideal Cutting Force Model

First of all, the cutter is divided into a finite number of disks with thickness equivalents to dl along the axial direction. Then, attach a static coordinate system $O_cX_cY_cZ_c$, which is different from the machine coordinate system (MCS), to the cutter and name it as the cutter coordinate system (CCS) in which the prediction of force will be accomplished. In this coordinate system, X_c is the feed direction of cutter and Z_c is attached to the axis of the cutter. In the machining, each disk is simplified as conducting oblique cutting. When the cutter is removing material from a workpiece, the force imposed on the disk could be divided into three components, including tangential force F_t, radial force F_r and axial force F_a, as shown in Fig. 3.6. These three forces on the jth disk of ith tooth could be expressed as:

$$\begin{cases} dF_t = K_{tc}h(\varphi)dl + K_{te}dl \\ dF_r = K_{rc}h(\varphi)dl + K_{re}dl \\ dF_a = K_{ac}h(\varphi)dl + K_{ae}dl \end{cases} \tag{3.39}$$

where K_{tc}, K_{rc}, K_{ac} are tangential, radial and axial shearing coefficients, K_{te}, K_{re}, K_{ae} are tangential, radial and axial edge coefficients, $h(\varphi)$ is the instantaneous undeformed chip thickness (IUCT). It can be approximately expressed as equation using feed peer tooth f_z:

$$h(\varphi) = f_z \sin\varphi \tag{3.40}$$

By transforming cutting forces into the static coordinate system, yields:

$$\begin{aligned} dF_{xc} &= -\frac{K_{tc}dl}{2} f_z \sin(2\varphi) - K_{te}dl\cos(\varphi) - \frac{K_{rc}dl}{2} f_z(1 - \cos(2\varphi)) - K_{re}dl\sin(\varphi) \\ dF_{yc} &= \frac{K_{tc}dl}{2} f_z(1 - \cos(2\varphi)) + K_{te}dl\sin(\varphi) - \frac{K_{rc}dl}{2} f_z \sin(2\varphi) + K_{re}dl\cos(\varphi) \\ dF_{zc} &= K_{ac}dlf_z\sin(\varphi) + K_{ae}dl \end{aligned} \tag{3.41}$$

At a particular time t, the forces imposed on a whole flute are the production of all disks in contact with workpiece. That is, the summation of all disks along axial direction could give us the force on the whole cutter:

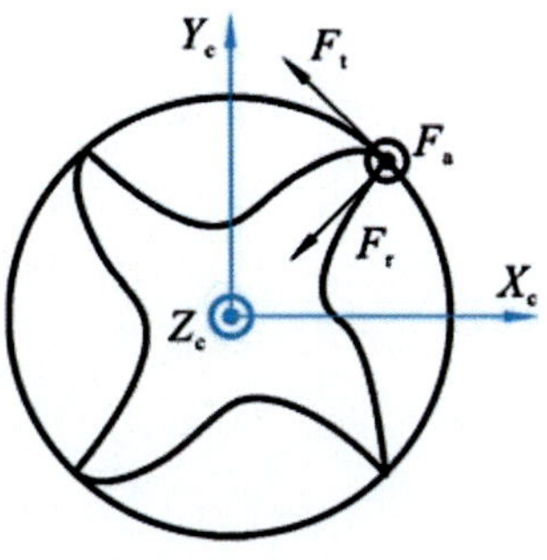

Fig. 3.6 Cutting forces components on one disk

$$\begin{cases} F_{xc} = -\frac{K_{tc}}{4k_\beta} f_z\cos(2\varphi) + \frac{K_{te}}{k_\beta}\sin\varphi + \frac{K_{rc}}{4k_\beta} f_z(2\varphi - \sin(2\varphi)) - \frac{K_{re}}{k_\beta}\cos\varphi \Big|_{\varphi_1}^{\varphi_2} \\ F_{yc} = -\frac{K_{tc}}{4k_\beta} f_z(2\varphi - \sin(2\varphi)) + \frac{K_{te}}{k_\beta}\cos\varphi - \frac{K_{rc}}{4k_\beta} f\cos(2\varphi) + \frac{K_{re}}{k_\beta}\sin\varphi \Big|_{\varphi_1}^{\varphi_2} \\ F_{zc} = \frac{K_{ac}}{k_\beta} f_z\cos\varphi - \frac{K_{ae}}{k_\beta}\varphi \Big|_{\varphi_1}^{\varphi_2} \end{cases} \tag{3.42}$$

where φ_2 and φ_1 are the upper and lower bounds of the integral, and they need to be judged because the actual contact status of cutting edges vary with the rotation of the cutter. Let l_1 represent the height from the point where the integration starts to the bottom of the tool, and l_2 represent the height from the point where the integration ends to the bottom of the tool $\left(l_1 \geq 0, l_2 \leq a_p\right)$. When the cutting edge does not contact with the workpiece, $l_1 = 0, l_2 = 0$ and $\varphi_1 = \varphi_2 = 0$. When the cutting edge contacts with the workpiece, it can be divided into the following two situations [8], as shown in Fig. 3.7.

(1) if $\frac{2tan\beta}{D} a_p \leq |\varphi_{st} - \varphi_{ex}|$, then

① when $\varphi_{st} < \varphi < \varphi_{st} + \frac{2tan\beta}{D} a_p$, then $l_1 = 0, l_2 \neq 0, l_2 \neq a_p$, and $\varphi_1 = \varphi, \varphi_2 = \varphi - \frac{2tan\beta}{D} l_2$.
② when $\varphi_{st} + \frac{2tan\beta}{D} a_p \leq \varphi \leq \varphi_{ex}$, then $l_1 = 0, l_2 = a_p$, and $\varphi_1 = \varphi, \varphi_2 = \varphi - \frac{2tan\beta}{D} a_p$.
③ when $\varphi_{ex} < \varphi < \varphi_{ex} + \frac{2tan\beta}{D} a_p$, then $l_1 \neq 0, l_2 \neq 0$, and $\varphi_1 = \varphi_{ex}, \varphi_2 = \varphi - \frac{2tan\beta}{D} a_p$.

(2) if $\frac{2tan\beta}{D} a_p > |\varphi_{st} - \varphi_{ex}|$, then

① when $\varphi_{st} < \varphi < \varphi_{ex}$, then $l_1 = 0, l_2 \neq 0, l_2 \neq a_p$, and $\varphi_1 = \varphi, \varphi_2 = \varphi - \frac{2tan\beta}{D} l_2$.
② when $\varphi_{ex} \leq \varphi \leq \varphi_{st} + \frac{2tan\beta}{D} a_p, l_1 \neq 0, l_2 \neq 0$, and $\varphi_1 = \varphi_{ex}, \varphi_2 = \varphi_{st}$.

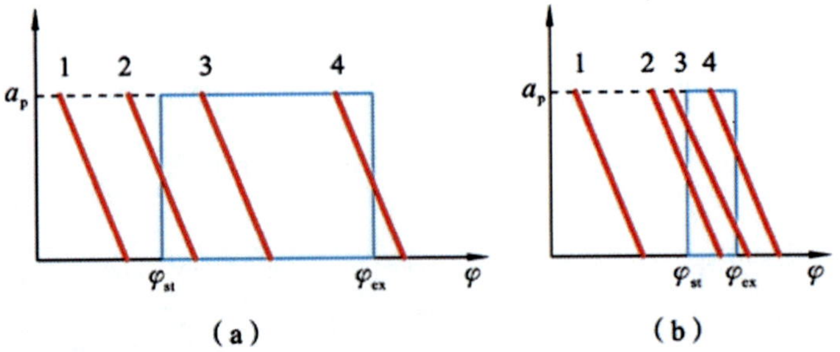

Fig. 3.7 Cutter contact situations

③when $\varphi_{st} + \frac{2tan\beta}{D}a_p < \varphi < \varphi_{ex} + \frac{2tan\beta}{D}a_p$, $l_1 \neq 0, l_2 = a_p$, and $\varphi_1 = \varphi_{ex}, \varphi_2 = \varphi - \frac{2tan\beta}{D}a_p$.

Introducing the sign function, the following equation can be derived as:

$$\begin{cases} \varphi_1 = \left(\varphi - \frac{2tan\beta}{D}l_1\right)\max\{sign(l_1), sign(l_2)\} \\ \varphi_2 = \left(\varphi - \frac{2tan\beta}{D}l_2\right)\max\{sign(l_1), sign(l_2)\} \end{cases} \tag{3.43}$$

where l_1 and l_2 can be calculated by cutter-workpiece geometric contact algorithm, φ_1 and φ_2 have nothing to do with the position angle of cutter φ. Therefore, this equation can be used for all types of machining except for slot milling. Then, l_1 and l_2 are functions of machining time, Eq. (3.43) can be rewritten as:

$$\begin{cases} \varphi_1(t) = \left[\varphi(t) - \frac{2\tan\beta}{D}l_1(t)\right]\max\{sign[l_1(t)], sign[l_2(t)]\} \\ \varphi_2(t) = \left[\varphi(t) - \frac{2\tan\beta}{D}l_2(t)\right]\max\{sign[l_1(t)], sign[l_2(t)]\} \\ \varphi(t) = \omega t \end{cases} \tag{3.44}$$

The calculation here requires the contact algorithm to discretely give the values of $l_1(t)$ and $l_2(t)$ at multiple time points $t_1, t_2, ..., t_n$, and these values constitute a matrix $\boldsymbol{L}$. If there are multiple cutting edges, many matrices are needed so that the cutting force of a specific cutting edge i can be written:

$$\begin{cases} F_{xc,i}(t) = \frac{K_{tc}}{4k_\beta}f(t)\left[-\cos\left(2\varphi_{2,i}(t)\right) + \cos\left(2\varphi_{1,i}(t)\right)\right] \\ \quad +\frac{K_{te}}{k_\beta}\left[\sin\left(\varphi_{2,i}(t)\right) - \sin\left(\varphi_{1,i}(t)\right)\right] \\ \quad +\frac{K_{rc}}{4k_\beta}f(t)\left[2\varphi_{2,i}(t) - \sin\left(2\varphi_{2,i}(t)\right) - 2\varphi_{1,i}(t) + \sin\left(2\varphi_{1,i}(t)\right)\right] \\ \quad +\frac{K_{re}}{k_\beta}\left[-\cos\left(\varphi_{2,i}(t)\right) + \cos\left(\varphi_{1,i}(t)\right)\right] \\ F_{yc,i}(t) = \frac{K_{tc}}{4k_\beta}f(t)\left[-2\varphi_{2,i}(t) + \sin\left(2\varphi_{2,i}(t)\right) + 2\varphi_{1,i}(t) - \sin\left(2\varphi_{1,i}(t)\right)\right] \\ \quad +\frac{K_{te}}{k_\beta}\left[\cos\left(\varphi_{2,i}(t)\right) - \cos\left(\varphi_{1,i}(t)\right)\right] \\ \quad +\frac{K_{rc}}{4k_\beta}f(t)\left[-\cos\left(2\varphi_{2,i}(t)\right) + \cos\left(2\varphi_{1,i}(t)\right)\right] \\ \quad +\frac{K_{re}}{k_\beta}\left[\sin\left(\varphi_{2,i}(t)\right) - \sin\left(\varphi_{1,i}(t)\right)\right] \\ F_{zc,i}(t) = \frac{K_{ac}}{k_\beta}f(t)\left[\cos\left(\varphi_{2,i}(t)\right) - \cos\left(\varphi_{1,i}(t)\right)\right] + \frac{K_{ae}}{k_\beta}\left[-\varphi_{2,i}(t) + \varphi_{1,i}(t)\right] \end{cases} \tag{3.45}$$

$$\begin{cases} \boldsymbol{L}_i = \begin{bmatrix} L_{1,i}(t) \\ L_{2,i}(t) \end{bmatrix} \\ \varphi_{1,i}(t) = \left[\omega t - \frac{2\tan\beta}{D}L_{1,i}(t)\right]\max\{sign[l_1(t)], sign[l_2(t)]\} \\ \varphi_{2,i}(t) = \left[\omega t - \frac{2\tan\beta}{D}L_{2,i}(t)\right]\max\{sign[l_1(t)], sign[l_2(t)]\} \\ t = t_1, t_2, ..., t_n \end{cases} \tag{3.46}$$

When multiple edges are in contact, the cutting force can be expressed as:

$$\begin{cases} F_{xc}(t) = \sum_{i=1}^{N} F_{xc,i}(t) \\ F_{yc}(t) = \sum_{i=1}^{N} F_{yc,i}(t) \\ F_{zc}(t) = \sum_{i=1}^{N} F_{zc,i}(t) \end{cases} \tag{3.47}$$

3.5.1.2 Undeformed Chip Thickness and Cutting Force Under Vibrational Condition

In the actual machining process, because of the limited stiffness of cutter-spindle subsystem, the dynamic displacement produced by cutter vibration is exactly inevitable. The real cutter-spindle subsystem can be regarded as a complex beam system with multiple cross-sections and variable shapes. The system is constrained by complicated support conditions (bearings and machine tool structure), and the coupling relationship between its internal parts also significantly affects the overall dynamic characteristics of the subsystem. The dynamic excitation of the system in the actual machining process is the cutting force on the tool tip. Due to the limited rigidity of the system and the complicated internal composition and boundary conditions, the vibration of the whole system is a continuous process in space and time. Theoretically, as a continuous system, the cutter-spindle subsystem vibrates under the action of cutting force, however, only the vibration operated on the contact area between the tool tip and the workpiece will produce wavy surface on the workpiece and affect the generation of cutting force. Thus, vibrations of other parts of the whole system on the cutting process can be ignored. The entire continuous vibration system can be equivalently simplified into a point vibration system composed of the tool tip. At the same time, the cutter-spindle subsystem has greater rigidity in the vertical direction, but the vibrations along the tool axis and Z_c direction can be ignored. Based on multi-step equivalent simplification, the continuous cutter-spindle subsystem can be simplified into a two-degree-of-freedom (2-DOF) vibration system, as shown in Fig. 3.8. In the cutter coordinate system, the vibration of this two-degree-of-freedom system can be expressed by a set of differential equations:

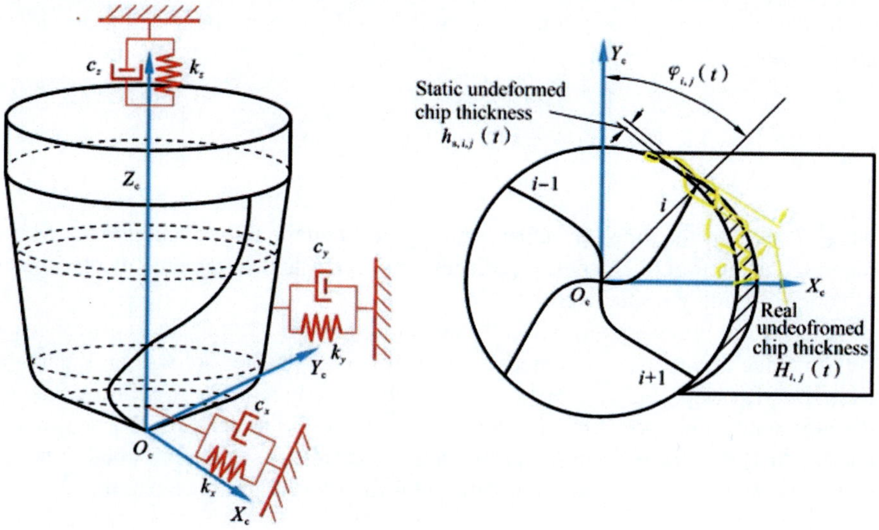

Fig. 3.8 Dynamic model and chip thickness of the cutter subsystem

$$\boldsymbol{M}\begin{bmatrix}\ddot{x}(t)\\ \ddot{y}(t)\end{bmatrix}+\boldsymbol{C}\begin{bmatrix}\dot{x}(t)\\ \dot{y}(t)\end{bmatrix}+\boldsymbol{K}\begin{bmatrix}x(t)\\ y(t)\end{bmatrix}=\begin{bmatrix}F_x(t)\\ F_y(t)\end{bmatrix} \tag{3.48}$$

where $\boldsymbol{M}$ is the modal mass matrix, $\boldsymbol{C}$ is the modal damping matrix, and $\boldsymbol{K}$ is the modal stiffness matrix.

Ignore the coupling in the x and y direction, this system can be expressed as:

$$\boldsymbol{M}\ddot{\boldsymbol{x}}(\mathrm{t})+\boldsymbol{C}\dot{\boldsymbol{x}}(\mathrm{t})+\boldsymbol{K}\boldsymbol{x}(t)=\mathbf{F}(t) \tag{3.49}$$

where $\boldsymbol{M}=\begin{bmatrix}m_x & 0\\ 0 & m_y\end{bmatrix}$, $\boldsymbol{C}=\begin{bmatrix}c_x & 0\\ 0 & c_y\end{bmatrix}$, $\boldsymbol{K}=\begin{bmatrix}k_x & 0\\ 0 & k_y\end{bmatrix}$, $\boldsymbol{F}(t)=\begin{bmatrix}F_x(t)\\ F_y(t)\end{bmatrix}$, $\boldsymbol{x}(t)=\begin{bmatrix}x(t)\\ y(t)\end{bmatrix}$. Except for the force and displacement, other dynamic parameters can be obtained by the modal test.

The vibration displacement bringing variation to the position of cutter that changes the Undeformed Chip Thickness (UCT) of cutting directly, thus influences the dynamic changes of cutting forces. In Fig. 3.8, the machined surface free of vibration is represented by the black line. The load of the cutter under free of vibration is determined by quasi-static UCT, which is denoted by $h_{s,i,j}(\varphi_{i,j}(t),\kappa_{i,j})$. When the cutter is vibrating, the flute will leave a wave on the newly machined surface. Temporarily, the UCT in machining progress is effected not only by the instantaneous displacement, but also the wave left on the external surface of chip by the former tooth that has removed material from workpiece. To take this influence into account in the calculation of cutting forces, an additional feed could be added to the ideal feed per tooth:

$$\boldsymbol{f}_v(t) = \boldsymbol{x}(t) - \boldsymbol{x}(t-T) = \begin{bmatrix} x(t) - x(t-T) \\ y(t) - y(t-T) \end{bmatrix} \tag{3.50}$$

where $T = \frac{\varphi_P}{\omega}$ is the time period between two consequential flutes cut into the workpiece, φ_P is the tool pitch angle of adjacent tooth, $x(t)$ and $y(t)$ are the displacement of cutter in x and y respectively.

Another situation to consider is that when the amplitude of vibration reaches a certain value and exceeds the theoretical feed per tooth, the tool will be separated from the workpiece, and the uncut material will directly increase the cutting load of the next tooth. One possible situation is more than one tooth leave the workpiece continuously when the vibration reaches a high magnitude. Under this condition, the additional feed on tooth is relevant with the former in-cut tooth's location:

$$x_P = \begin{bmatrix} x_P \\ y_P \end{bmatrix} = \begin{bmatrix} \min\{x(t-T), x(t-2T), \cdots, x(t-KT)\} \\ \min\{y(t-T), y(t-2T), \cdots, y(t-KT)\} \end{bmatrix} \tag{3.51}$$

So that, additional feed peer tooth that causes dynamic variation of UCT is expressed as:

$$\boldsymbol{f}_v(t) = \boldsymbol{x}(t) - \boldsymbol{x}_P(t) = \begin{bmatrix} \Delta x(t) \\ \Delta y(t) \end{bmatrix} = \begin{bmatrix} x(t) - x_P \\ y(t) - y_P \end{bmatrix} \tag{3.52}$$

The $\boldsymbol{f}_v$ in Eq. (3.52) realizes the time-delay coupling effect considering the continuous cutting of the tooth, and eliminates the time-delay coupling term in the tool vibration equation. Finally, the UCT on the jth disk of the ith flute is represented as:

$$h_{d,i,j}(\varphi_{i,j}(t), \kappa_{i,j}) = \left[\sin(\varphi_{i,j}(t))\sin(\kappa_{i,j}) \ \cos(\varphi_{i,j}(t))\sin(\kappa_{i,j})\right]\boldsymbol{f}_v(t) \tag{3.53}$$

where $\kappa_{i,j}$ is the axial immersion angle.

$ds_{i,j}$ is the length for the jth disk of the ith flute, and $db_{i,j}$ is the chip thickness for the jth disk of the ith flute. According to the dynamic UCT, the cutting forces can be written as:

$$\begin{bmatrix} F_{t,i,j}(t) \\ F_{r,i,j}(t) \\ F_{a,i,j}(t) \end{bmatrix} = g(\varphi_{i,j}(t)) \begin{bmatrix} K_{tc}H(\varphi_{i,j}(t), \kappa_{i,j})db_{i,j} + K_{te}ds_{i,j} \\ K_{rc}H(\varphi_{i,j}(t), \kappa_{i,j})db_{i,j} + K_{re}ds_{i,j} \\ K_{ac}H(\varphi_{i,j}(t), \kappa_{i,j})db_{i,j} + K_{ae}ds_{i,j} \end{bmatrix}$$

$$H_{i,j}(t) = h_{s,i,j}(\varphi_{i,j}(t), \kappa_{i,j}) + h_{d,i,j}(\varphi_{i,j}(t), \kappa_{i,j}) \tag{3.54}$$

where $H_{i,j}(t)$ is the actual UCT acts on the rake face. Then, cutting forces on the whole cutter are the sum of forces on all incut disks:

$$\begin{bmatrix} F_x(t) \\ F_y(t) \\ F_z(t) \end{bmatrix} = \sum_{i=1}^{N}\sum_{j=1}^{M} g\big(\varphi_{i,j}(t)\big)\boldsymbol{T}(\varphi_{i,j}(t), z_j)\begin{bmatrix} F_{t,i,j}(t) \\ F_{r,i,j}(t) \\ F_{a,i,j}(t) \end{bmatrix} \tag{3.55}$$

Due to the coupling effect of vibration and cutting forces, the prediction procedure is a nonlinear procedure. Particularly, the solution of system dynamic equations is accomplished by applying classical forth order Runge–Kutta method with assumption of zero initial condition (both the initial values of velocity and displacement are zero). Meanwhile, the time step in simulation is same to the one used in experiments that is determined by sampling frequency. The complete flow of the prediction as follows:

(1) At each time instant t_k, the rotatory angle of cutter (φ) is identified with the number of disks (M) in contact and entry ($\varphi_{j,st}(t_k)$) and exit ($\varphi_{j,ex}(t_k)$) angle for each disk.
(2) At the same time, the previous position of cutter ($\boldsymbol{x}_P(t_k)$) removed material from workpiece corresponding to a current time instant is obtained.
(3) Current cutter position ($\boldsymbol{x}(t_k)$) is calculated and the forces ($F_{t,i,j}(t_k)$, $F_{r,i,j}(t_k)$, $F_{a,i,j}(t_k)$) on each disk is calculated after acquiring position angle ($\varphi_{i,j}(t_k)$) of each disk and whether one disk is in contact or not by judging $g(\varphi_{t,i,j}(t_k))$.
(4) Both real-time cutting forces ($F_x(t_k)$, $F_y(t_k)$, $F_z(t_k)$) and cutter displacement ($\boldsymbol{x}(t_k)$, $\boldsymbol{y}(t_k)$) and velocity ($\dot{\boldsymbol{x}}(t_k)$, $\dot{\boldsymbol{y}}(t_k)$) would then be stored to be used in later predictions.
(5) If all the prediction instants are completed, an algorithm is terminated and results are exported. Otherwise, return the (1) step.

Different with traditional methods summating of forces on each disk, this procedure has the indispensable section of solving cutter vibration following the calculation of forces (Step (4)), which accounts for a large part of the whole program. Before the calculation of forces on each disk, the previous cutter position engaged into the workpiece must be selected from stored data (Step (2)).

When the axial depth of cut a_p is small enough (empirically about 1 ~ 2 mm), a_p could be equivalent to the length of the thickness dz of a disk. In the meanwhile, the radial depth of cut a_e in many common processes is also small enough to permit only one tooth in contact with workpiece at an arbitrary moment. This circumstance is relatively normal in machining titanium alloy or high-temperature alloy. Then the cutting force model could be greatly simplified:

$$\begin{bmatrix} F_x(t) \\ F_y(t) \\ F_z(t) \end{bmatrix} = \sum_{i=1}^{N} g(\varphi_i(t)) \boldsymbol{T}(\varphi_i(t), z) \begin{bmatrix} K_{tc} H_i(t)db + K_{te}ds \\ K_{rc} H_i(t)db + K_{re}ds \\ K_{ac} H_i(t)db + K_{ae}ds \end{bmatrix} \tag{3.56}$$

where

$$\boldsymbol{T}\left(\varphi_{i,j}(t), z_j\right) = \begin{bmatrix} -\cos\left(\varphi_{i,j}(t)\right) & -\sin\left(\varphi_{i,j}(t)\right)\sin\left(\kappa_{i,j}\right) & -\sin\left(\varphi_{i,j}(t)\right)cos\left(\kappa_{i,j}\right) \\ \sin\left(\varphi_{i,j}(t)\right) & -\cos\left(\varphi_{i,j}(t)\right)\sin\left(\kappa_{i,j}\right) & -\cos\left(\varphi_{i,j}(t)\right)\sin\left(\kappa_{i,j}\right) \\ 0 & -\cos\left(\varphi_{i,j}(t)\right) & -\sin\left(\varphi_{i,j}(t)\right) \end{bmatrix}$$

At this time, the contact calculation only requires the entry angle φ_{st}, exit angle φ_{ex} and position angle φ. When $\varphi_{st} \leq \varphi \leq \varphi_{ex}$, $\delta = 1$, otherwise, $\delta = 0$.

3.5.2 *Cutting Force Coefficients Identification Considering Vibration*

Many scholars take into account the vibration conditions in the machining process, so that the dynamic response of the process system can be predicted more realistically and accurately. At the same time, this kind of consideration of the actual machining conditions also provides a way for the identification of cutting force coefficients closer to the real machining conditions [9].

3.5.2.1 Cutting Force Identification Model

During machining, the force at time moment t_i can be obtained directly, thus establishing a correspondence:

$$\begin{bmatrix} F_{tc}(t_i) \\ F_{rc}(t_i) \\ F_{ac}(t_i) \end{bmatrix} = \begin{bmatrix} \left[K_{tc}h_{ac}(t_i) + K_{te}\right]a_p \\ \left[K_{rc}h_{ac}(t_i) + K_{re}\right]a_p \\ \left[K_{ac}h_{ac}(t_i) + K_{ae}\right]a_p \end{bmatrix} = \begin{bmatrix} -\cos\varphi(t_i) & -\sin\varphi(t_i) & 0 \\ \sin\varphi(t_i) & -\cos\varphi(t_i) & 0 \\ 0 & 0 & 1 \end{bmatrix}^{-1} \begin{bmatrix} F_{xt}(t_i) \\ F_{yt}(t_i) \\ F_{zt}(t_i) \end{bmatrix} \tag{3.57}$$

In the above equation, there are six unknown variables $K_{tc}, K_{te}, K_{rc}, K_{re}, K_{ac}, K_{ae}$, so at least six equations are needed to solve the equations, and the forces on one point are not enough to solve it. Suppose there are n collected data in the period when the cutter rotate one revolution, when the data sampling frequency is high enough, $3n > 6$ is easy to be achieved. Thus a set of equations can be obtained:

$$\begin{bmatrix} \begin{bmatrix} K_{tc}h_{ac}(\varphi(t_1)) + K_{te} \\ K_{tc}h_{ac}(\varphi(t_2)) + K_{te} \\ \vdots \\ K_{tc}h_{ac}(\varphi(t_n)) + K_{te} \end{bmatrix}_{n\times 1} \\ \begin{bmatrix} K_{rc}h_{ac}(\varphi(t_1)) + K_{re} \\ K_{rc}h_{ac}(\varphi(t_2)) + K_{re} \\ \vdots \\ K_{rc}h_{ac}(\varphi(t_n)) + K_{re} \end{bmatrix}_{n\times 1} \\ \begin{bmatrix} K_{ac}h_{ac}(\varphi(t_1)) + K_{ae} \\ K_{ac}h_{ac}(\varphi(t_2)) + K_{ae} \\ \vdots \\ K_{ac}h_{ac}(\varphi(t_n)) + K_{ae} \end{bmatrix}_{n\times 1} \end{bmatrix} = \begin{bmatrix} \frac{1}{a_p}\begin{bmatrix} F_{tc}(t_1) \\ F_{tc}(t_2) \\ \vdots \\ F_{tc}(t_n) \end{bmatrix}_{n\times 1} \\ \frac{1}{a_p}\begin{bmatrix} F_{rc}(t_1) \\ F_{rc}(t_2) \\ \vdots \\ F_{rc}(t_n) \end{bmatrix}_{n\times 1} \\ \frac{1}{a_p}\begin{bmatrix} F_{ac}(t_1) \\ F_{ac}(t_2) \\ \vdots \\ F_{ac}(t_n) \end{bmatrix}_{n\times 1} \end{bmatrix} \tag{3.58}$$

Rewrite to:

$$\begin{bmatrix} \boldsymbol{M} & \boldsymbol{0} & \boldsymbol{0} \\ \boldsymbol{0} & \boldsymbol{M} & \boldsymbol{0} \\ \boldsymbol{0} & \boldsymbol{O} & \boldsymbol{M} \end{bmatrix}_{3n\times 6} \begin{bmatrix} K_{tc} \\ K_{te} \\ K_{rc} \\ K_{re} \\ K_{ac} \\ K_{ae} \end{bmatrix} = \begin{bmatrix} \begin{bmatrix} F_{tc}(t_1) \\ F_{tc}(t_2) \\ \vdots \\ F_{tc}(t_n) \end{bmatrix} \\ \begin{bmatrix} F_{rc}(t_1) \\ F_{rc}(t_2) \\ \vdots \\ F_{rc}(t_n) \end{bmatrix} \\ \begin{bmatrix} F_{ac}(t_1) \\ F_{ac}(t_2) \\ \vdots \\ F_{ac}(t_n) \end{bmatrix} \end{bmatrix}_{3n\times 1} \tag{3.59}$$

Eq. (3.59) are typical regular equations and can be solved using least square method. After a simple matrix operation, and let $\boldsymbol{A} = \boldsymbol{M}^T\boldsymbol{M}$, $\boldsymbol{B}_P = \boldsymbol{M}^T \left[F_{pc}(t_1)\ F_{pc}(t_2) \ldots F_{pc}(t_n)\right]$ $(p = t, r, a)$, then:

$$\begin{bmatrix} \boldsymbol{A} & \boldsymbol{0} & \boldsymbol{0} \\ \boldsymbol{0} & \boldsymbol{A} & \boldsymbol{0} \\ \boldsymbol{0} & \boldsymbol{0} & \boldsymbol{A} \end{bmatrix} \begin{bmatrix} K_{tc} \\ K_{te} \\ K_{rc} \\ K_{re} \\ K_{ac} \\ K_{ae} \end{bmatrix} = \begin{bmatrix} \boldsymbol{B}_t \\ \boldsymbol{B}_r \\ \boldsymbol{B}_a \end{bmatrix} \tag{3.60}$$

where,
$A = \begin{bmatrix} \sum_{i=1}^{n} h_{ac}(t_i)^2 & \sum_{i=1}^{n} h_{ac}(t_i) \\ \sum_{i=1}^{n} h_{ac}(t_i) & n \end{bmatrix}$, $B_p = \begin{bmatrix} \sum_{i=1}^{n} h_{ac}(t_i) F_{pc}(t_i) \\ \sum_{i=1}^{n} F_{pc}(t_i) \end{bmatrix}$.
Then the solution is:

$$\begin{bmatrix} K_{tc} \\ K_{te} \\ K_{rc} \\ K_{re} \\ K_{ac} \\ K_{ae} \end{bmatrix} = \begin{bmatrix} \boldsymbol{A} & \boldsymbol{0} & \boldsymbol{0} \\ \boldsymbol{0} & \boldsymbol{A} & \boldsymbol{0} \\ \boldsymbol{0} & \boldsymbol{0} & \boldsymbol{A} \end{bmatrix}^{-1} \begin{bmatrix} \boldsymbol{B}_t \\ \boldsymbol{B}_r \\ \boldsymbol{B}_a \end{bmatrix} \tag{3.61}$$

3.5.2.2 Cutter Displacement Calculation

The tool's two-degree-of-freedom vibration model can be expressed directly by Eq. 3.48, and when the real-time change cutting force is obtained, the real-time displacement of the tool can be obtained by solving the difference equation system. Here is the differential equation group:

$$\begin{cases} m_x x''(t) + c_x x'(t) + k_x x(t) = F_x(t) \\ m_y y''(t) + c_y y'(t) + k_y y(t) = F_y(t) \end{cases} \tag{3.62}$$

Make the following substitute:

$$\begin{cases} x_1(t) = x'(t) \\ x_1(t_0) = x'(t_0) = x_1,\ x(t_0) = x_0 \\ y_1(t) = y'(t) \\ y_1(t_0) = y(t_0) = y_1,\ y(t_0) = y_0 \end{cases} \tag{3.63}$$

The above question can be translated into the solving of following equation groups:

$$\begin{cases} m_x x_1'(t) + c_x x_1(t) + k_x x(t) = F_x(t) \\ m_y y_1'(t) + c_y y_1(t) + k_y y(t) = F_y(t) \end{cases} \tag{3.64}$$

Combine the four-level, four-order classical Runge–Kutta formula:

$$\begin{cases} y_{n+1} = y_n + \frac{h}{6}(K_1 + 2K_2 + 2K_3 + K_4) \\ K_1 = f(x_n, y_n) \\ K_2 = f\left(x_n + \frac{1}{2}h,\ y_n + \frac{h}{2}K_1\right) \\ K_3 = f\left(x_n + \frac{1}{2}h,\ y_n + \frac{h}{2}K_2\right) \\ K_4 = f(x_n + h,\ y_n + hK_3) \end{cases} \tag{3.65}$$

Finally the displacement in x direction can be derived as:

$$x_{k+1} = x_k + \frac{h}{6}(K_1 + 2K_2 + 2K_3 + K_4) \tag{3.66}$$

$$x_{1,k+1} = x_{1,k} + \frac{h}{6}(L_1 + 2L_2 + 2L_3 + L_4) \tag{3.67}$$

Next,

$$\begin{cases} K_1 = x_{1,k} \\ K_2 = x_{1,k} + \frac{h}{2}L_1 \\ K_3 = x_{1,k} + \frac{h}{2}L_2 \\ K_4 = x_{1,k} + \frac{h}{2}L_3 \end{cases} \tag{3.68}$$

$$\begin{cases} L_1 = f(t_k, x_k, x_{1,k}) = -\frac{c_x}{m_x}x_{1,k} - \frac{k_x}{m_x}x_k + \frac{F_x(t_k)}{m_x} \\ L_2 = f\left(t_k + \frac{h}{2}, x_k + \frac{h}{2}K_1, x_{1,k} + \frac{h}{2}L_1\right) = -\frac{c_x}{m_x}\left(x_{1,k} + \frac{h}{2}L_1\right) - \frac{k_x}{m_x}\left(x_k + \frac{h}{2}K_1\right) + \frac{F_x\left(t_k + \frac{h}{2}\right)}{m_x} \\ L_3 = f\left(t_k + \frac{h}{2}, x_k + \frac{h}{2}K_2, x_{1,k} + \frac{h}{2}L_2\right) = -\frac{c_x}{m_x}\left(x_{1,k} + \frac{h}{2}L_2\right) - \frac{k_x}{m_x}\left(x_k + \frac{h}{2}K_2\right) + \frac{F_x\left(t_k + \frac{h}{2}\right)}{m_x} \\ L_4 = f(t_k + h, x_k + hK_3, x_{1,k} + hL_3) = -\frac{c_x}{m_x}(x_{1,k} + hL_3) - \frac{k_x}{m_x}(x_k + hK_3) + \frac{F_x(t_k + h)}{m_x} \end{cases} \tag{3.69}$$

The same can be said y direction displacement. By taking the initial displacement and the velocity to zero, you can obtain the vibration displacement for a certain period of time. Combined with the first two sections, the cutting force coefficient for each cycle of the tool can be calculated.

References

1. HOU Y F. Detection and monitoring recognition and learning optimization method of time-varying factors in milling condition [D]. Xi'an; Northwestern Polytechnical University, 2015.
2. ALTINTAS Y. Manufacturing automation: metal cutting mechanics, machine tool vibration, and CNC design [M]. 2nd. Cambridge, UK: Cambridge University Press, 2012.
3. TEITENBERG T M, BAYOUMI A E, YUCESAN G. Tool wear modeling through an analytic mechanistic model of milling processes [J]. Wear, 1992, 154(2): 287–304.
4. LAPSLEY J T, GRASSI R C, THOMSEN E G. Correlation of plastic deformation during metal cutting with tensile properties of the work material [J]. Transactions of the ASME, 1950, 72(7): 979–986.
5. WALDORF D J. Shearing, ploughing, and wear in orthogonal machining [D]. Urbana: University of Illinois at Urbana-Champaign, 1996.
6. KAPOOR S G, DEVOR R E, ZHU R, et al. Development of mechanistic models for the prediction of machining performance: model building methodology [J]. Machining Science and Technology, 1998, 2(2): 213–238.
7. SMITHEY D W, KAPOOR S G, DEVOR R E. A new mechanistic model for predicting worn tool cutting forces [J]. Machining Science and Technology, 2001, 5(1): 23–42.

8. YAO Q, LUO M, ZHANG D H, et al. Identification of cutting force coefficients in machining process considering cutter vibration [J]. Mechanical Systems and Signal Processing, 2018, 103: 39–59.
9. YAO Q. Cutting force coefficient identification method considering dynamic properties effect from heterogeneous process systems [D]. Xi'an; Northwestern Polytechnical University, 2019.

Chapter 4
Learning and Optimization of Process Model

In the current study, the modeling and analysis for cutting process and machining process system both abstracted and simplified the system. Although it could describe the machining process and machining system in this way, the response of the system cannot be accurately predicted because of the complexity of the actual process and system. Relative characters are embodied in the following aspects:

(1) In the actual machining process, tool wear, surface state and other factors caused by tool manufacturing error and material batch difference are significantly different from the experimental results of specimens. The machining parameters window calculated by the theoretical model is not fully applicable to the actual situation, and the prediction results are also remarkably distinct from the actual machining tool wear and system vibration.
(2) The integral blisk, annular casing and other parts in aero-engines have complex structures. The contact and cooling state, machine tool pose and other process conditions between the tool and the workpiece are constantly changing. Thus, the data under steady cutting conditions in the laboratory are not fully suitable for the variable working circumstances.
(3) The actual machining process of complex parts is often affected by some fluctuations or interference factors, such as hard points in the workpiece material, cooling pressure fluctuations and uncertain interference factors. The prediction model usually does not consider such influences, thereby, it cannot effectively predict and control the manufacturing defects caused by fluctuation or interference factors.

Therefore, in view of the actual processing requirements for the key components of aero-engines, it is necessary to combine the rapidly developing sensing, monitoring technologies and data processing methods, use the data collected on-site for learning, modify the theoretical model constantly according to the actual situation, and optimize the machining process, to achieve high-performance manufacturing for key components. For example, in the process of tool wear identification, in order

D. Zhang et al., *Intelligent Machining of Complex Aviation Components*, Research on Intelligent Manufacturing, https://doi.org/10.1007/978-981-16-1586-3_4

to realize tool wear condition evaluation based on monitoring data, Monte Carlo Method can be used to establish the probability mapping relationship between wear correlation coefficients and different wear state of cutting edges according to cutting test data.

4.1 Learning and Optimization Method of the Machining Process Model

At the scene of actual parts processing, experienced operators can determine whether the processing is reasonable according to their observed machining status information, such as noise, the surface texture of the workpiece, etc., and then optimize the machining sequence, process parameters or clamping mode of the workpiece. At the same time, they can also judge whether the machining compensation or correction is performed. Before the batch production of new parts, by means of the above trial cutting process, the process and parameters are continuously improved, and finally, be confirmed.

The learning optimization for process model refers to the above-mentioned manual trial cutting method, which changes the manual observation to the sensor monitoring, and replaces optimization based on the human experience to the machine. As shown in Fig. 4.1, for a new part in the intelligent machining system, the optimal process parameters are generally given by combining the empirical data and simulation optimization method. Then, the sensor is used to monitor the actual machining process, and the mapping between the monitoring data and the processing sequence and position of parts is established, so as to accurately analyze the problems existing in the process. On this basis, the set optimization model is adopted to optimize the machining process, and the optimization results are utilized in the next product. Furthermore, the above method is continuously applied to optimize other products. Finally, through the repeated application of the above procedures, the process model and parameters which describe the specific workpiece processing more precisely and stably can be obtained.

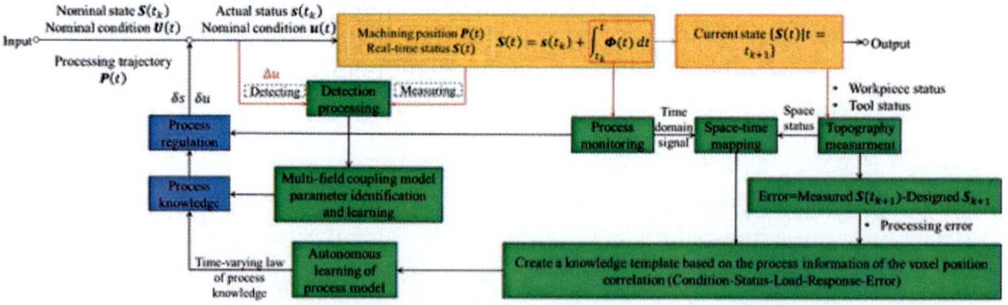

Fig. 4.1 The general process of the learning and optimization method of the process model

The above learning and optimization method can be applied to many machining scenes, including the optimization of process parameters, determination of process system model parameters, confirmation of deformation error compensation coefficient, etc. The iteration of the above process can make the process model more useful and accurate.

4.2 Time-Position Mapping of Processing Data

There are a lot of on-site monitoring data in the intelligent machining system, which play an important role in analyzing the geometric model of the workpiece or the evolution of dynamic characteristics and tool wear. These data are also the basis of machining process optimization. Therefore, in intelligent machining, it is necessary to conduct time-position mapping, so as to correspond the machining position and time sequence of the workpiece to process data [1].

As shown in Fig. 4.2, firstly, the CLS file can be obtained by NC programming of the workpiece. Then, the CLS file can get G code through post builder. At last, G code executes the motion of the machine tool motion. Each line of data in the CLS file can match with each line of G code, and also corresponds to the specific position

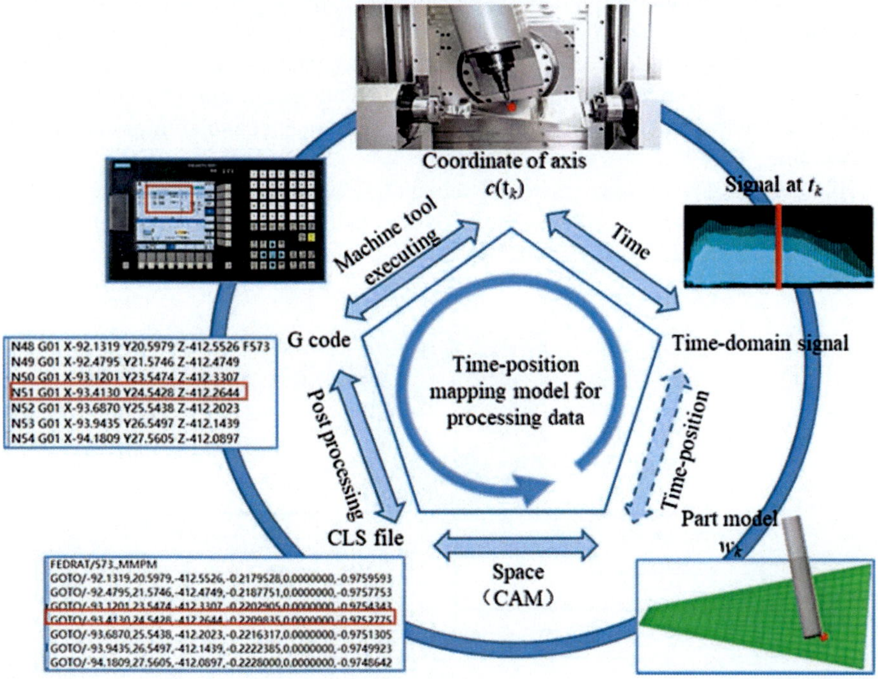

Fig. 4.2 Time-position mapping method for process data

on the part. When the NC machine tool executes the G code, the real-time processing data, including the tool spindle subsystem and the workpiece fixture subsystem, can be recorded. That is to say, the monitoring data can also accurately correspond to the instructions in the G code.

According to the above mapping relation, the monitoring signal can be accurately mapped with the specific position of the workpiece model surface, so as to establish the time-position mapping relation of the processing data.

Taking the annular casing of the aero-engine as an example, the whole process of the casing is discretized in time. Each spatially discrete depth voxel of the process system has corresponding processing information, thus, the spatial mapping relationship between processing information and depth voxel can be established [1].

For the complex machining process of annular thin-walled parts, the space subdivision method can be used to simplify the processing. The depth voxels divided along the *U* and *V* directions of the workpiece surface are taken as the material removal unit. Then, each depth voxel connected in sequence along the toolpath constitutes the material removal sequence. Combined with the structural dynamic modification algorithm, this method can calculate the dynamic evolution of the process system along the material removal sequence. At the same time, optimizing the cutting parameter for the single-row toolpath can ensure stable cutting in the material removal sequence. And the information related to the material removal sequence, that is, the machining process, including the predicted values of the process system characteristics such as the natural frequency and the normal stiffness of the process system, and the initial values of the process parameter optimization such as the spindle speed, the axial depth of cut, etc., under the steady cutting state needs to be stored in the corresponding depth voxels in a certain format. The results can provide basic data for online control and learning evolution of the subsequent annular thin-walled parts processing.

Figure 4.3 shows a space mapping schematic diagram of the machining process information in the current step. For a certain moment *t* in the material removal sequence, the depth voxel *E* corresponds to it. The workpiece surface parameter

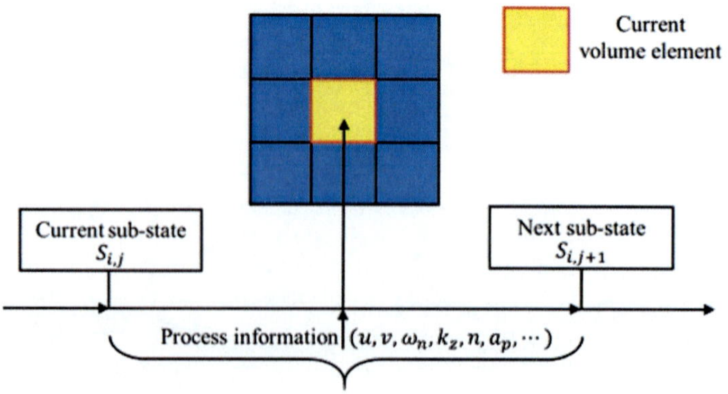

Fig. 4.3 The space mapping of the machining process information in the current step

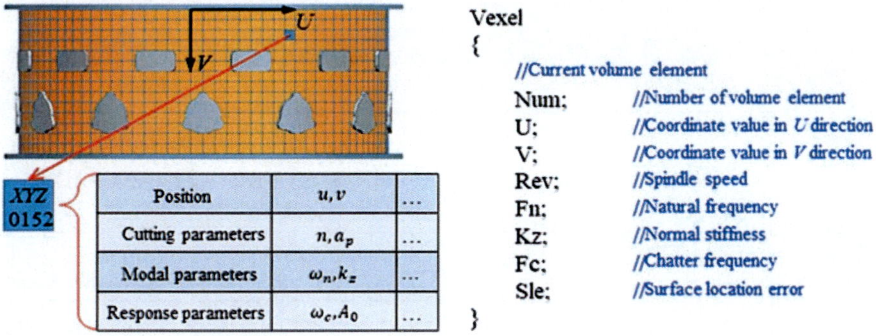

Fig. 4.4 Recording of process knowledge

coordinates corresponding to the current cutting position are u and v, the corresponding natural frequency of the process system is w_n, and the normal stiffness is k_z. The spindle speed that meets the stable machining conditions corresponding to the cutting row of the voxel is n, and the axial depth of cut is a_p. This information needs to be recorded in accordance with the given storage format.

The machining process information corresponding to each depth voxel is stored along the material removal sequence to form a machining process optimization scheme for the dynamic evolution prediction and dynamic response control of the annular thin-walled parts milling process. Machining process information can be classified and stored according to voxel position, cutting parameters, modal and response characteristics, as shown in Fig. 4.4. All information constitutes the process knowledge for the machining of annular thin-walled parts. The control coefficients and cutting parameters of the process model can be continuously refined by means of knowledge mining, learning, evolution, etc. And finally, stable and reliable process knowledge is formed and recorded in the database, in order to guide actual production, improve the processing quality and efficiency of parts.

For the entire machining process of the casing from the current processing sub-state $\boldsymbol{S}_{i,j}$ to the next processing sub-state $\boldsymbol{S}_{i,j+1}$, the spatial mapping between the geometric parameters, physical parameters, process parameters of the machining process and each depth voxel can be established according to the material removal sequence. The information of these parameters is stored on each depth voxel to provide initial prediction values for the subsequent online control of the actual machining process.

Figure 4.5 shows the spatial mapping of machining process information corresponding to cutting line l_5 in the material cutting scheme of a certain section of casing parts, and the detailed machining process information is shown in Table 4.1. In order to facilitate the description of the spatial mapping process, the outer surface of the casing is expanded in a plane, and the material removal units in each cutting row can sequentially correspond to the horizontal axis of the cutting segment sequence. The cutting segment $l_{5,1}$ and $l_{5,7}$ corresponding to workpiece surface parameter positions are (0.042, 0.393) and (0.542, 0.393), the natural frequencies are 1411 Hz and

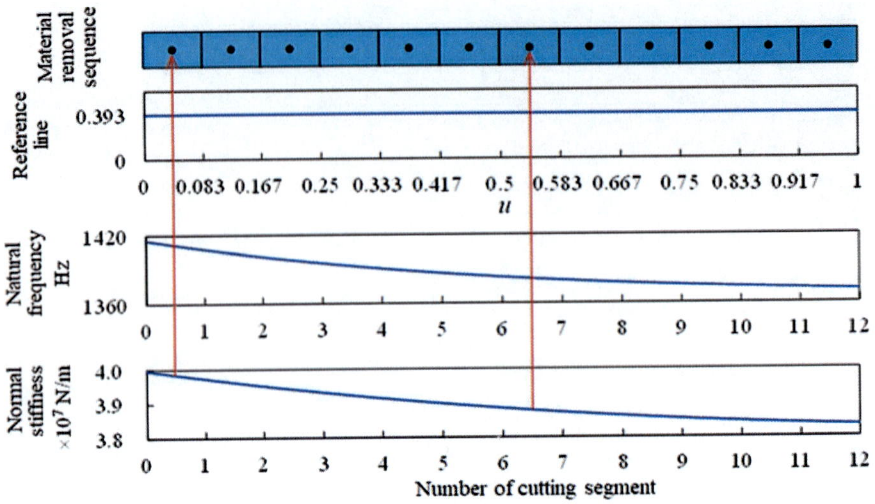

Fig. 4.5 The spatial mapping of machining process information corresponding to cutting row l_5

1380 Hz, the normal stiffnesses are 3.973×10^7 N/m and 3.873×10^7 N/m, the spindle speeds are 3237 r/min and 3173 r/min, the fault-tolerant cutting depths are 0.685 mm and 0.672 mm, and the absolute stable cutting depths are 0.149 mm and 0.145 mm. It should be noted that the cutting segment $l_{5,0}$ indicates that the cutting row l_4 has completed the material removal, whereas the cutting row l_5 has not yet started material removal.

According to the above method, the spatial mapping machining process information is performed on all cutting rows of the outer surface of the casing, then the predicted values of the process system characteristics and the optimized initial values of the process parameters along the material removal sequence in the entire process can be obtained.

Figure 4.6a shows the normal stiffness distribution corresponding to the evolution of the workpiece geometry. In order to facilitate the data analysis, the ring surface of the casing is expanded along the u direction. It can be found that the normal stiffness at the top of the casing is smaller than that of the root of the casing, while the normal stiffness at the middle of the casing which is near the collar of the boss 1 is the smallest. And the normal rigidity near the root of the casing increases significantly. For a single cutting row, the material removal of each cutting segment will cause changes in the dynamic characteristics of the process system, and lead to different results of cutting parameter optimization at each cutting segment.

Figure 4.6b depicts the spindle speed distribution corresponding to the cutting row l_5 of the casing processing. Because of the existence of the boss structure on the cutting row l_5, for fear of collision interference, it is necessary to perform repeated

Table 4.1 Machining process information corresponding to cutting row l_5

Number of cutting segment	Geometric position (u, v)	Natural frequency (Hz)	Normal stiffness ($\times 10^7$ N/m)	Spindle speed (rpm)	Tolerance cutting depth (mm)	Absolutely stable depth (mm)
$l_{5,0}$	(0.042, 0.393)	1415	3.997	3244	0.688	0.150
$l_{5,1}$	(0.042, 0.393)	1411	3.973	3237	0.685	0.149
$l_{5,2}$	(0.125, 0.393)	1402	3.951	3217	0.681	0.148
$l_{5,3}$	(0.208, 0.393)	1395	3.932	3203	0.680	0.147
$l_{5,4}$	(0.292, 0.393)	1390	3.916	3192	0.677	0.146
$l_{5,5}$	(0.375, 0.393)	1386	3.899	3184	0.675	0.146
$l_{5,6}$	(0.458, 0.393)	1382	3.885	3176	0.673	0.145
$l_{5,7}$	(0.542, 0.393)	1380	3.873	3173	0.672	0.145
$l_{5,8}$	(0.625, 0.393)	1378	3.861	3169	0.670	0.144
$l_{5,9}$	(0.708, 0.393)	1376	3.853	3165	0.669	0.144
$l_{5,10}$	(0.792, 0.393)	1374	3.847	3161	0.669	0.144
$l_{5,11}$	(0.875, 0.393)	1372	3.839	3157	0.668	0.144
$l_{5,12}$	(0.958, 0.393)	1371	3.834	3155	0.667	0.143

feeding and retracting operations between the boss when planning toolpaths. Therefore, it is possible to optimize a spindle speed at each cutting segment and change the spindle speed during the process of feeding and retracting. At the same time, since the change amplitude of the dynamic characteristic of the process system is small in each cutting row of the casing processing, a spindle speed can be optimized for a single cutting row.

Figure 4.6c describes the spindle speed distribution corresponding to each cutting row of the casing process. It can be found that each cutting row corresponds to an optimized spindle speed which could ensure steady cutting of this row.

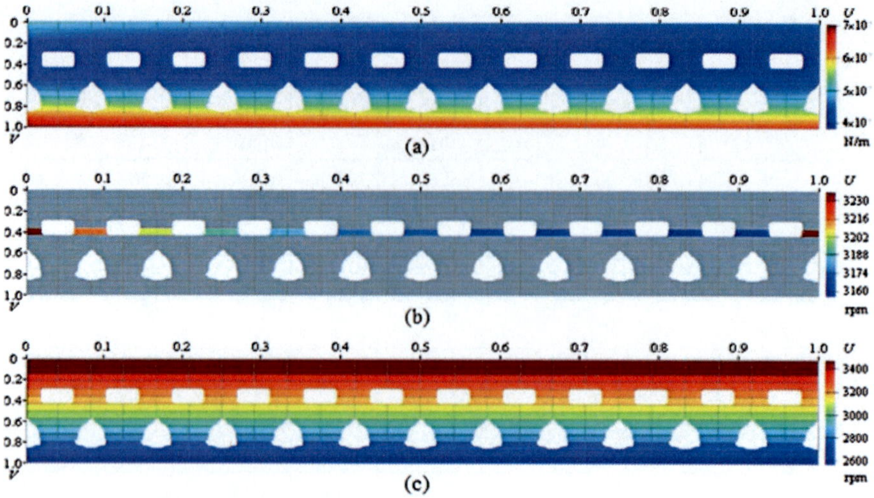

Fig. 4.6 Spatial mapping results of machining process information **a** the normal stiffness distribution corresponding to the evolution of the workpiece geometry. **b** the spindle speed distribution corresponding to the cutting row l_5. **c** the spindle speed distribution corresponding to each cutting row

4.3 Iterative Learning Method of Machining Error Compensation

Machining is a complex elastoplastic deformation process with the generation and transformation of cutting force and heat. In the meantime, the existence of different ways of clamping, variation of stiffness and instability of machine tool result in unavoidable cutting vibrations and deformation and affect machining accuracy and surface quality. Specifically, the machining process is a functional process f with inputs and outputs. The inputs include the conditions of the machine tool, cutter, fixture, workpiece, cutting parameters (cutting depth, spindle speed and feed rate) and other independent variables. The outputs are the status of the workpiece after cutting. When machining thin-walled parts, affected by the cutting force, the deformation of parts always makes the machining accuracy drop significantly. In the case that the clamping method cannot be optimized and the weak rigidity characteristic cannot be changed, choosing a suitable cutting depth can effectively offset the error caused by the deformation of the knife. Therefore, how to efficiently optimize the depth of cut is a problem faced by thin-walled parts machining [2].

4.3.1 *In-Position Detection Method for Workpiece Geometry Information*

In order to realize the error compensation, it is necessary to detect the workpiece error during the machining process. There are various methods to measure the data of complex surfaces. As shown in Fig. 4.7, according to whether the probe contacts the surface of the measured object, it is divided into contact type and non-contact type.

The contact measurement method can be subdivided into point trigger type and continuous scanning type, and refers to that it records the coordinate position of the product surface point through the contact between the sensor probe and the product. Contact measuring instruments include inductance meters and a Coordinate Measuring Machine (CMM), among which CMM is the main one. It has high measurement accuracy for the surface data of rigid objects and is primarily used for measuring machined workpieces and detecting geometric dimensions and tolerances. With the advent of analog probes, CMM can continuously scan and measure the blade profile. At present, CMM is widely used in the actual inspection of aero-engine blades. The advantages of the contact measurement method are: (1) high accuracy; (2) rapidly measuring basic geometry without being affected by the environment, material, and color of the entity; (3) measuring dead angles such as deep grooves and small grooves that cannot be achieved by optical instruments.

The non-contact measurement method is based on the basic principles of optics, acoustics, magnetism and other fields, and converts a certain amount of physical analog into coordinate points on the product surface through appropriate algorithms. At present, there are mainly optical scanning, CT scanning (X-ray, γ-ray), and ultrasonic methods. Optical scanning measurement is divided into laser point measurement, linear measurement, area measurement and incandescent lamp surface grating measurement. Optical scanning can be divided into laser optical scanning and incandescent raster optical scanning according to different light sources used. The laser point scanning method is generally the same as CMM measuring method, but the

Fig. 4.7 Classification of data measurement methods

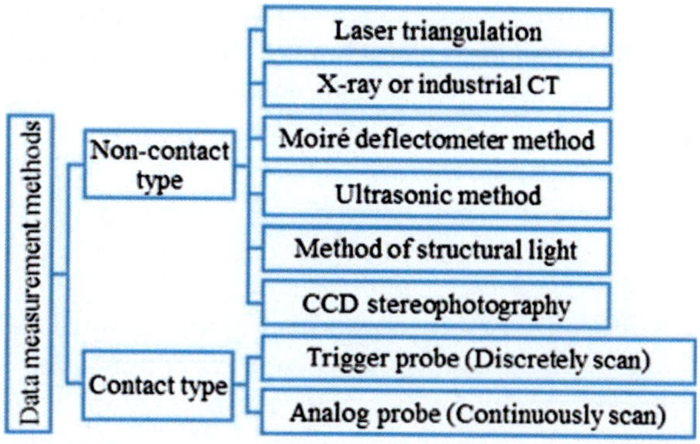

probe is different. The optical line or area scanning is based on the optical triangulation principle, that is, the measuring light source with regular geometric shapes is projected onto the measured surface to form a diffuse reflect light band, which is imaged on the Charge-coupled Device (CCD), and measures the spatial coordinate of a certain position on the measured surface according to the triangulation principle. Industrial CT adopts the attenuation law and distribution of X-ray which is in the detected object to obtain detailed information of the object by the detector display, and finally displays it in the form of the image by applying computer information processing and image reconstruction technology. The advantages of the non-contact measurement method are: (1) fast measurement speed; (2) high resolution; (3) measuring most of the features on the object; (4) no requirement for probe radius compensation; (5) measuring objects that cannot be measured the contact measurement method, such as soft objects, plastics thin pieces, etc. Currently, this measuring method is generally used in product digitization and reverse engineering. Its shortcomings are: (1) poor measurement accuracy; (2) easily affected by external environmental factors; (3) more difficult to complete the measurement of geometric shapes, such as slender and deep holes. Although contact measurement is less efficient than non-contact measurement, it is widely used in CAD/CAM, product inspection and quality control of the manufacturing industry because of its high accuracy and intelligence and low equipment price.

In recent years, the development of on-machine contact measurement technology and non-contact optical measurement technology has provided reliable data sources for compensation of processing errors and quality inspection between workpiece processing procedures.

4.3.2 *Compensation Modeling of Machining Errors for Thin-Wall Parts*

If the cutting force is too large in the machining of thin-walled parts, the tool will bend and deform because of small tool stiffness. In thin-walled parts processing, although the stiffness of the material to be processed is large, the small thickness makes the material exhibit low rigidity. In this case, the influence of tool deformation can be ignored. Therefore, the deformation of the tool during the cutting process can be represented by the deformation of the workpiece [3]:

$$\varepsilon = \frac{F}{K} \tag{4.1}$$

where ε is the deformation value; F is the normal cutting force; K is the stiffness.

In this deformation, the modeling of cutting force is more complicated, especially when the depth of cut is changed, the variation of cutting force is not nonlinear. In

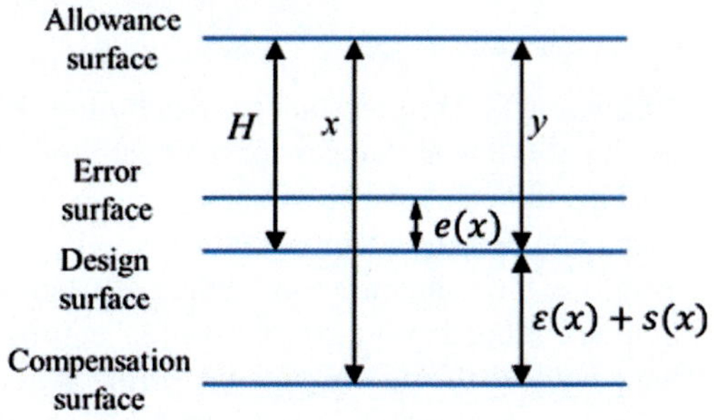

Fig. 4.8 The relationship between the parameters in the compensation process

addition, the stiffness K is related to the clamping method and thickness distribution of the entire workpiece. Particularly for the dynamic process of machining, the stiffness of each area on the workpiece does not decrease linearly when the depth of cut increases. Therefore, when only considering elastic deformation, the amount of elastic deformation varying with cutting depth does not change linearly. What's more, in view of the influence of inelastic errors in the machining system, the nonlinearity between machining error and cutting depth is more obvious, and can be expressed as:

$$e(x) = \varepsilon(x) + s(x) - x + H \tag{4.2}$$

where x is nominal cutting depth; e is machining error; s is inelastic error; H is design allowance. The relationship between the parameters is shown in Fig. 4.8.

On the basis of analyzing the causes of tool deformation and machining error and their nonlinear relationship with cutting depth, a general model of error compensation for thin-walled parts is established by combining the compensation method. The process is described as follows:

For thin-walled parts processing, the design allowance H is set to constant before the machining. The nominal cutting depth is x, the real cutting depth is y, and the machining error is e. When other independent variables are fixed and the stiffness parameter is ignored, the machining process f is only related to x and y, which can be expressed as:

$$y = f(x) \tag{4.3}$$

Compared with Eq. (4.2), the machining error can be expressed as:

$$e(x) = H - f(x) \tag{4.4}$$

Equation (4.4) indicates that the final error is related to the cutting process. Combined with the analyzing of deformation for thin-walled parts, the machining process is expressed as:

$$f(x) = x - \varepsilon(x) - s(x) \tag{4.5}$$

Equation (4.5) shows that the machining process f is a non-linear process, which leads to non-linear changes for the machining error. These non-linear function relationships indicate that for a specific cutting process, the machining error e is related to the choice of the nominal cutting depth x and these two variables are non-linear. Therefore, the compensation method of machining error is selecting a more suitable nominal cutting depth x to make the real cutting depth y equal to the design allowance H and achieve the purpose of zero error.

The iteration method can be introduced to solve the nonlinear equation as a successive approximation approach. The iterative error compensation equation, which is the general error compensation model, is established to calculate the nominal cutting depth of the kth cutting process as:

$$\begin{cases} x_0 = H \\ x_{k+1} = x_k + \rho_{k+1} \cdot e_k, \ (k = 0, 1, 2, \ldots) \end{cases} \tag{4.6}$$

where ρ is the compensation coefficient, k is the times of the compensation process. The case that $k = 0$ corresponding to the condition that the part is machined without compensation.

4.3.3 *Solution of Error Compensation Model for Thin-Walled Parts*

The purpose of error compensation is to reduce the machining error to fit the accuracy range. For the cutting error model, error compensation is developed to calculate the x when $e = 0$, which is the zero-point problem of $e(x)$. Combined with the previous analysis, the calculation of the error compensation can be expressed as:

$$f(x) = f(x_0) + f'(x_0)\Delta x + \frac{1}{2} f''(x_0)\Delta x^2 \tag{4.7}$$

Taking different values for different items in the above formula, there are four different zero-point iteration calculation methods: Mirror Method, Newton Iteration Method, Previous Point Secant Method, and Initial Point Secant Method.

4.3.3.1 Mirror Method

Whether the error prediction method or the error measurement method is used, the mirror compensation method is mostly considered when calculating the compensation amount. That is, the error generated after processing according to a certain

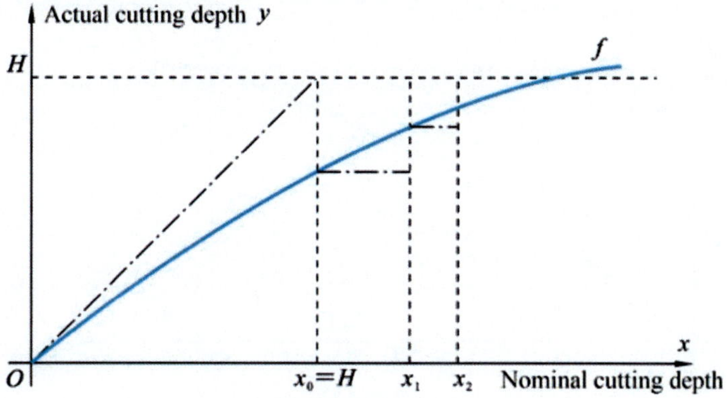

Fig. 4.9 The iterative process of the mirror method

nominal depth will be added to the nominal depth of the next machining with the same amount. It can be written as:

$$x_{k+1} = x_k + e_k \ (k = 0, 1, 2, \ldots) \tag{4.8}$$

It can be seen from Eq. (4.8) that no matter the relationship between the nominal cutting depth and the real cutting depth in the machining process is linear or non-linear, the error can be reduced by a series of iterative machining. It is also the simplicity and applicability of this method that makes it suitable for various error compensation scenarios. The iterative effect of the mirror method can be proved by compensation coefficients, and its iterative process is shown in Fig. 4.9. It can be seen that convergence speed is slow.

4.3.3.2 Newton's Method

Newton's method is an important numerical method for solving approximate values of the root of nonlinear equations. Its basic idea is to use linear parts of nonlinear equations to gradually approach the approximate value. The iteration process can be expressed as:

$$x_{k+1} = x_k + \frac{1}{y'_k} \cdot e_k \ (k = 0, 1, 2, \ldots) \tag{4.9}$$

Newton's method has a second-order convergence rate, and the iteration process is shown in Fig. 4.10.

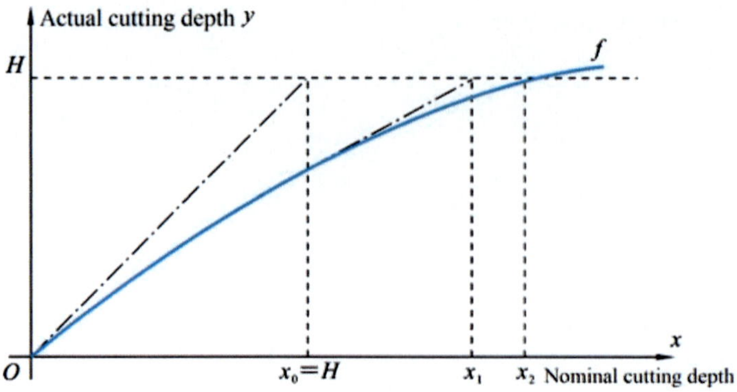

Fig. 4.10 Compensation iterative process for Newton's method

4.3.3.3 Previous Point Secant Method

Newton's method needs to calculate the first derivative value of the machining process f in each iterative computation. Due to the complexity of f, it's difficult to determine the relationship between the nominal cutting depth and the real cutting depth. This is also the reason why using the iterative method to solve the error model. However, this complex relationship makes it difficult to calculate the first derivative value of f, which makes Newton's method can only be expressed as a mathematical principle. In order to avoid calculating derivatives, $\frac{1}{y_k'}$ which is in the Newton iteration formula is replaced by the average rate of change $\frac{x_k - x_{k-1}}{y_k - y_{k-1}}$. Then Eq. (4.9) becomes the following form:

$$x_{k+1} = x_k + \frac{x_k - x_{k-1}}{y_k - y_{k-1}} \cdot e_k \ (k = 0, 1, 2, \ldots) \tag{4.10}$$

When calculating the average rate of change of each step, the information of the first two steps (x_k, y_k) and (x_{k-1}, y_{k-1}) is needed. This iterative method is known as the two-step method. When $k = 0$, let $x_{k-1} = y_{k-1} = 0$.

Using the method of replacing the derivative with the average rate of change, the geometric meaning of the iteration changes from the tangent method to the secant method. More specifically, the average rate of change utilizes the secant between the current point and the previous point, thus, this method is called the previous point secant method. The previous point secant method has a convergence rate of 1.618 order, and the iteration process is shown in Fig. 4.11.

4.3.3.4 Initial Point Secant Method

The aforementioned iterative formula construction methods are all based on the calculation zero points for nonlinear equations. Combined with the principle of elastic

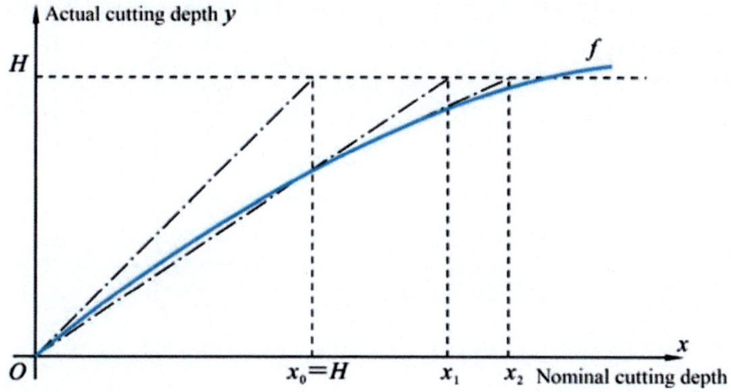

Fig. 4.11 Compensation iterative process for previous point secant method

deformation of thin-walled parts processing, the error can be controlled from the view of physical and geometrical aspects. On this basis, the quotient of deformation and nominal cutting depth, the springback coefficient λ is introduced and presented in the following:

$$\lambda = \frac{\varepsilon}{x} \tag{4.11}$$

For ε, it can be expressed as:

$$\varepsilon_k = e_k + \sum_{i=1}^{k} z_i \tag{4.12}$$

where z_i is the difference value of the ith processing nominal cutting depth and the i−1th processing nominal cutting depth.

As shown in Fig. 4.12, the uncompensated process can be achieved when $k = 0$, where the nominal cutting depth is x_0 and machining error is e_0, thus the springback coefficient could be calculated as:

Fig. 4.12 The relationship between the parameters in the uncompensated process

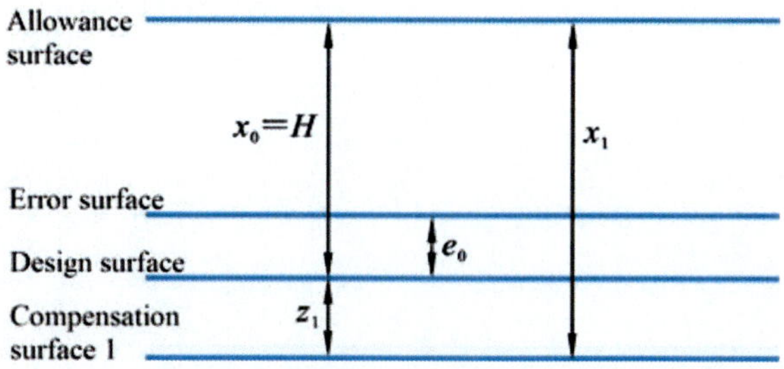

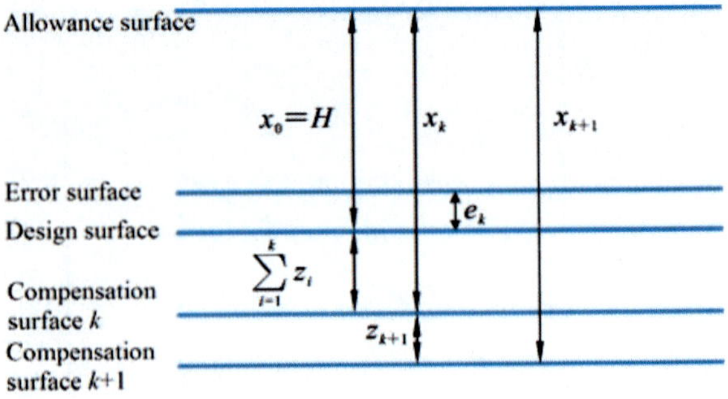

Fig. 4.13 The relationship between the parameters after the kth compensated process

$$\lambda_0 = \frac{e_0}{x_0} \tag{4.13}$$

Considering that the current stiffness and cutting force are fixed and utilizing the springback coefficient, the nominal cutting depth of the next compensated process, x_1, could be calculated under the control of the ideal error, zero. The relationship between the nominal cutting depth and increment of the next compensated process is expressed as:

$$(x_0 + z_1) \cdot \lambda_0 = z_1 \tag{4.14}$$

$$z_1 = \frac{x_0}{H - e_0} \cdot e_0 \tag{4.15}$$

$$x_1 = x_0 + z_1 \tag{4.16}$$

During the kth compensated process, the nominal cutting depth is x_k and the machining error is e_k, hence the springback coefficient λ_k could be calculated. And then, as shown in Fig. 4.13, considering that the current stiffness and cutting force are fixed and calculating the nominal cutting depth, x_{k+1}, of the $k + 1$th compensated process under the control of the ideal error, zero, the springback coefficient can be expressed as:

$$\lambda_k = \frac{e_k + \sum_{i=1}^{k} z_i}{x_k} \tag{4.17}$$

$$(x_k + z_{k+1}) \cdot \lambda_k = z_{k+1} + \sum_{i=1}^{k} z_i \tag{4.18}$$

$$z_{k+1} = \frac{x_k}{H - e_k} \cdot e_k \tag{4.19}$$

$$x_{k+1} = x_k + z_{k+1} \tag{4.20}$$

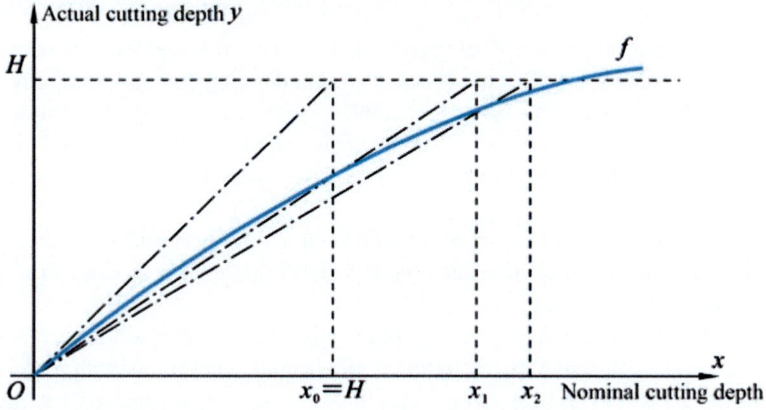

Fig. 4.14 Compensation iterative process for initial point secant method

After transformation, the error compensation equation can be obtained:

$$x_{k+1} = x_k + \frac{x_k}{y_k} \cdot e_k \ (k = 0, 1, 2, \ldots) \tag{4.21}$$

The iterative process of the initial point secant method is shown in Fig. 4.14, it has a one order convergence speed which is lower than that of the previous point secant method.

4.3.4 *Learning Control Method for Error Compensation Coefficient*

As shown in Table 4.2, due to the different calculation methods of the compensation

Table 4.2 Model solution comparison

Category	Mirror method	Newton's method	Previous point secant method	Initial point secant method
Compensation coefficient	1	$\frac{1}{y'_k}$	$\frac{x_k - x_{k-1}}{y_k - y_{k-1}}$	$\frac{x_k}{y_k}$
Computation difficulty	Easiest	Hardest	Hard	Easy
Convergence order	/	2	1.618	1
Convergence speed	Slower	Fastest	Faster	Fast

coefficient, the iterative convergence speed and computation difficulty of the aforementioned four solutions are distinguishing. Based on the previous secant method and elastic deformation, the initial point secant method establishes an iterative algorithm with clear physical meaning. Especially for the processing of multiple-parts-single-step parts, before the error is stable, each part not only represents an iterative machining process but also denotes the off-line calculation method of compensation coefficient.

As shown in Fig. 4.15, this section takes the machining of multiple-parts-single-step parts as an example to introduce the compensation coefficient learning control method based on the initial point secant method.

Multiple-parts single step refers to reasonably selecting machining method of current cutting depth, after batch parts completing all operations which are before the current operation. Single-step indicates that only one step is required, and multiple parts correspond to the selection of different depths of the cutting process. It can be expressed as:

$$\omega \in \{\omega_k | k \in \{0, 1, 2, \ldots\}\} \tag{4.22}$$

where ω_k indicates the kth part.

When the machining error meets the accuracy requirements, the machining reaches a stable state, and the compensation iteration stops. Then the remaining parts are expressed as:

$$\omega \in \{\omega_k | k \in \{n, n+1, \ldots\}\} \tag{4.23}$$

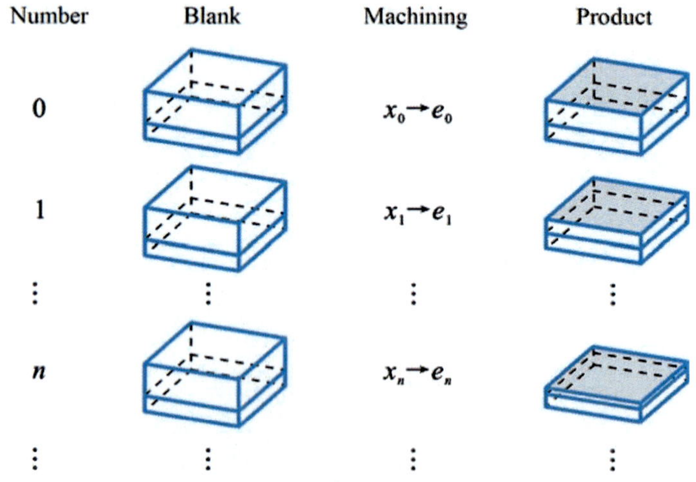

Fig. 4.15 Schematic diagram of multiple-parts-single-step machining

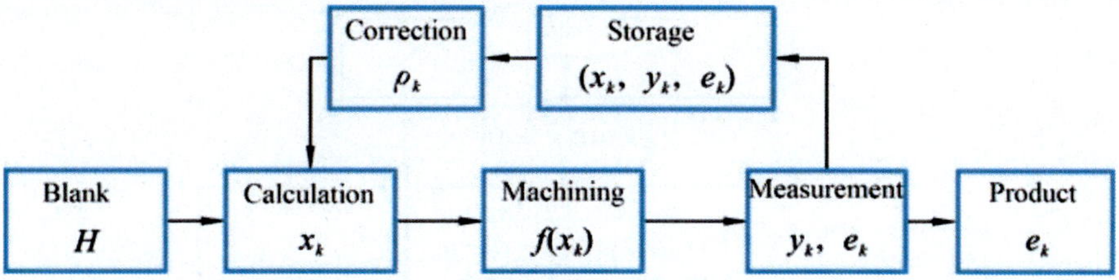

Fig. 4.16 Block diagram of compensation coefficients control

That is, the cutting depth of the remaining parts is x_n, until the machining of all parts finish.

The block diagram of compensation coefficients control is shown in Fig. 4.16. The control system uses the feedback control principle to correct and compensate the nominal cutting depth of test parts before the error is stable. The correction method refers to the initial point secant method to obtain the compensation coefficient for each processing, namely:

$$\rho_{k+1} = \frac{x_k}{y_k}, k = 0, 1, 2, \ldots \tag{4.24}$$

where x_k can be calculated by Eq. (4.6) and Eq. (4.21) and y_k can be obtained by measuring.

Equation (4.24) manifests that the control of the next machining process requires the previous processing state. Although this control principle cannot parallelly adjust the current machining process in real time, the next process can still be controlled by using the previous processing state to calculate the compensation coefficient offline, so as to achieve the purpose of discrete dynamic adjustment for multi-part processing. The flow chart is shown in Fig. 4.17, which presents that even if the parts shown in 4.23 may be out of tolerance during processing, the cutting depth of the next part can still be corrected through the feedback control of the compensation coefficient to ensure the stability of the system.

4.3.5 The Application of Error Iterative Compensation Method in Thin-Walled Blade Machining

The blades of aeroengines are typical thin-walled parts. The thinnest part of the leading and trailing edge is less than 0.3mm, which is easy to deform during the processing. The comparison of blade processing error before and after the application of the above-mentioned processing error iterative compensation method is shown in Fig. 4.18. Without the compensation, the machined concave surface is overcut while the convex surface is undercut (see Fig. 4.18a, c, respectively). Applying the

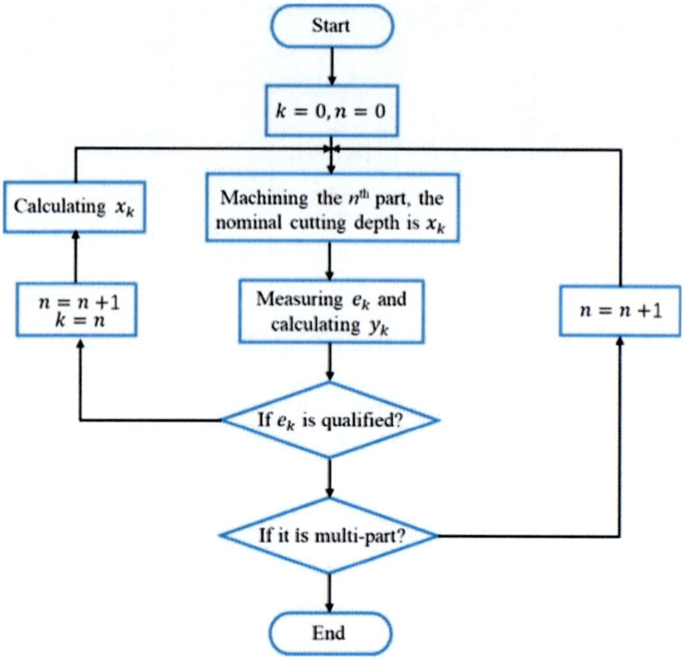

Fig. 4.17 Error compensation flow

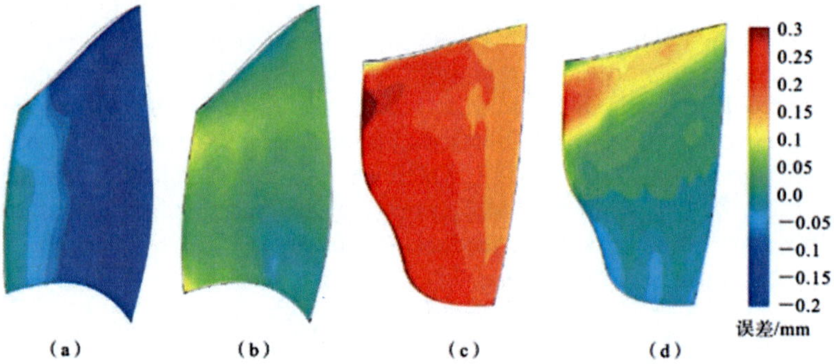

Fig. 4.18 The error distributions in a different area of blades **a** uncompensated concave surface **b** compensated concave surface. **c** uncompensated convex surface **d** compensated convex surface

compensated milling process, the overcut of the concave surface and undercut of convex surface are corrected shown in Fig. 4.18b, d. At the same time, the average errors of the concave surface are reduced to 25.1% while the average errors of the convex surface are reduced to 14.3%.

The error distributions of the leading and trailing edges are shown in Fig. 4.19. The

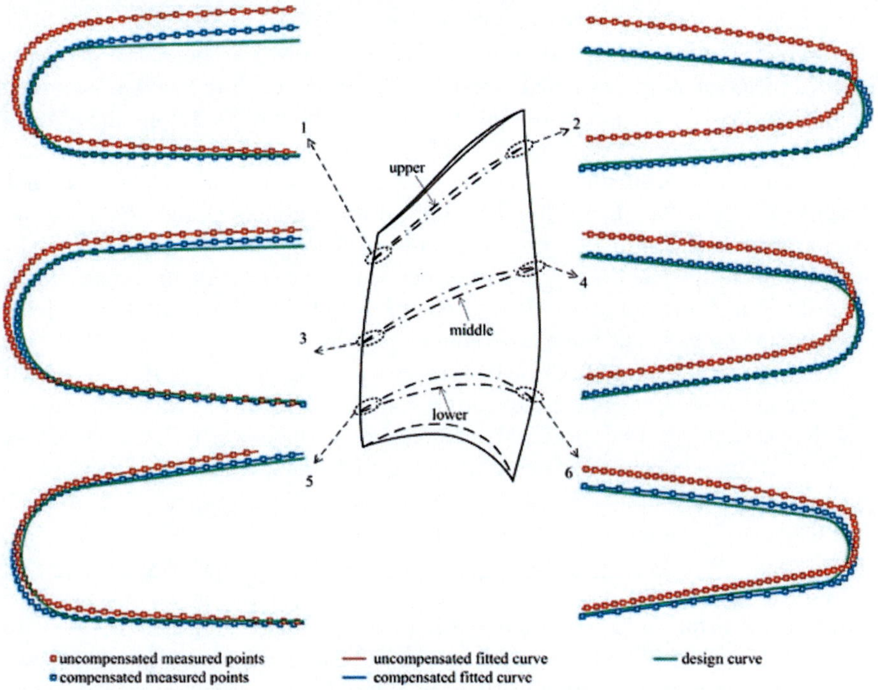

Fig. 4.19 Measured points of specific areas after compensating

red measured points represent the actual machining error distributions of the uncompensated blade which is deviated from the designed section curves (green points). After the compensation process, the blue measured points show that the compensated curves are much closer to the design curves, indicating that the overcut and undercut are compensated. The compensated average errors of leading and trailing edges are reduced to 29.7% of the uncompensated machining process.

4.4 Iterative Learning Optimization Method for Deep-Hole Drilling Depth

In the manufacturing industry, hole-making occupies approximately 30% of all metal cutting operations and is one of the significant processes in parts machining. Due to the flexibility and low cost of twist drills, the most common method of hole-making is to use twist drills to drill holes. The difficulty of drilling is chip evacuation. Compared with milling and turning, the chips generated during drilling are constrained by the hole wall and drill pipe chip flutes. The chips accumulate in the chip flutes because of narrow chip evacuation space and operate friction and pressure with drill flute and hole wall. As the drilling depth increases, the chips continue to accumulate, and the drilling force increases rapidly. Among all types of holes, the hole with its length to diameter ratio exceeding five is defined as the deep hole. The geometric characteristics of deep holes determine that deep-hole drilling has the features of small chip evacuation space and long chip evacuation path. Therefore, the problem of chip removal is particularly significant in deep-hole drilling. At the same time, the geometric characteristics of the large depth-to-diameter ratio of the deep hole directly determine that the drill pipe used for drilling is slender, the rigidity is poor and tool breakage is prone to occur when the drilling force is too large. Generally, hole-making is the final process of parts processing. Once the cutter is broken during the drilling, it is difficult to take it out from the deep hole, which will directly result in the scrapping parts and huge economic losses [4].

To avoid drill breakage because of excessive drilling force caused by chip accumulation in the process of deep-hole drilling, the peck drilling method is often used to drill in enterprises. The specific implementation method is as follows. The drill is fed to a certain predetermined depth in each drilling step, and then is retracted to a position above the workpiece surface with the coolant flushing the chip out from the drill flutes, and then again fed to drill a further particular depth in the next drilling step. This process continues until the nominal hole depth is achieved. During peck drilling, the drilling depth in each drilling step is a critical factor. If the value of the drilling depth is too large, it may cause drill breakage due to large drilling force; if it is too small, there will be too many retractions and toolpath length during the peck drilling process and the machining efficiency is low. In order to solve these problems, an effective method must be taken to find the optimal drilling depth in each drilling step, which maximizes the efficiency and avoids drill breakage.

In view of the shortcomings of existing deep-hole drilling depth optimization methods, the iterative learning optimization method can be used [5]. This method combines the advantages of modeling prediction and online monitoring. Firstly, the relationship between drilling force and drilling depth is established as a theoretical model for iterative learning. Then, according to the process characteristics of the peck drilling method, the drilling cycle process in real machining is carried out for iterative learning and modeling revision. Then the setting value of the drilling depth of the next iteration is given according to the iterative formula of the drilling depth to guide the drilling cycle process in real time and improve the processing efficiency under the premise of non-cutter broken. At the same time, after the completion of the iterative learning process, the maximum single drilling depth under the processing conditions is given as the reference value of the next drilling.

4.4.1 Chip Evacuation Force Model for One-Step Drilling

Since the drilling force in the deep-hole drilling process increases with the increase of drilling depth, in order to optimize the drilling depth, it is necessary to establish a quantitative relationship between the drilling force and the drilling depth in a single drilling process to predict the maximum drilling depth without tool broken.

In the drilling process, the cutter makes linear feed movement perpendicular to the workpiece plane. During the drill-in process, the drill is fed into the workpiece with the chisel edge first contacting the workpiece surface and extruding the material flowing to the cutting lips. Then the contact length between cutting lips and the workpiece material is gradually increasing, leading to the drilling force increasing because of the apex angle of the cutter. After the cutting lips completely immerse into the workpiece material, the drilling force stays stable because of the constant chip load, that is, the cutting area on the cutting lips due to the fixed feed rate and cutting speed. The drilling process is illustrated in Fig. 4.20. After that, the workpiece material removed by cutting lips is converted to chips flowing in the drill flutes. The chips generate pressure and friction when reacting with the drill flute the hole wall. The force subjected to the chips is denoted as the chip evacuation force. As the drilling depth increases, the generated chips gradually accumulate, resulting in the chip evacuation force and the drilling force gradually increasing.

The chip evacuation force and the force subjected to the drill flute by the chips are a pair of balance forces. The maximum force that the drill flute can withstand is the main constraint condition for the drilling depth. In the process of deep-hole drilling, the drilling force includes shear force and chip evacuation force. For the given spindle speed and feed speed, the cutting load is constant, thus, the shear force hardly changes during the drilling, except for the drill-in process. However, with the increase of drilling depth, the chip evacuation force as well as its proportion in the deep-hole drilling force gradually climbs, which has become the main factor restricting the depth of deep-hole drilling. Therefore, only researching the increasing stage of the chip evacuation force in the drilling process, the shear force component is subtracted from drilling force during data processing to obtain the chip evacuation force when using the model to express the relationship between the chip evacuation

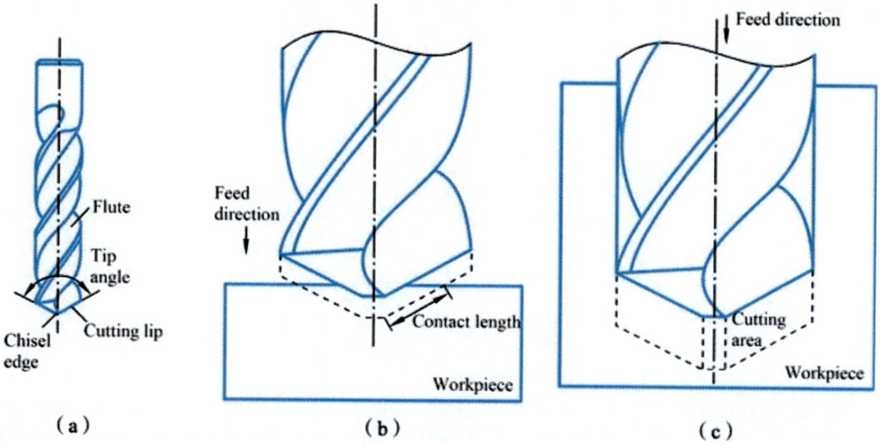

Fig. 4.20 Drill geometry and drilling process, **a** geometry of a typical twist drill **b** drill-in process.**c** drilling process after full immersion of the drill bit

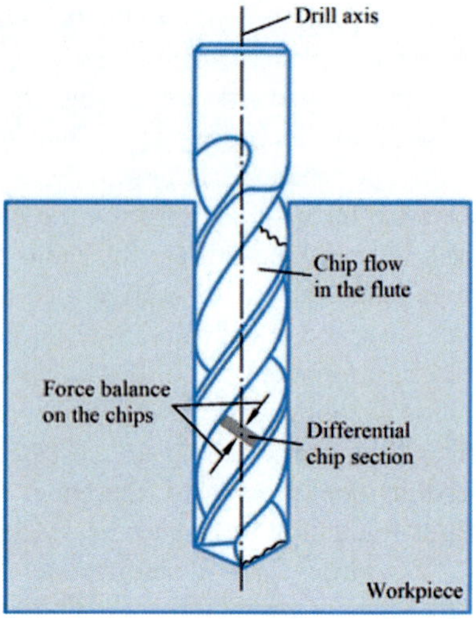

Fig. 4.21 Force analysis of differential chip in deep-hole drilling

force and the drilling depth. The general form of cutter broken in deep-hole drilling is torsion fracture, so the chip evacuation torque model can generally be used.

The movement of the chip flow in the flutes can be divided into two components: the translation along the flutes and the revolving around the drill axis. As shown in Fig. 4.21, discretize the chip into a number of differential chip elements along the flute direction, as for each differential chip element, it is under the force balance of the friction and pressure applied by the drill flute and the hole wall, as well as the force from the adjacent differential chip elements.

From this, the material mechanics parameters, friction coefficient, the geometric parameters of the chip flutes, etc. can be introduced to derive the relationship between the chip torque and the drilling depth. Chip evacuation torque generally varies exponentially with the change of drilling depth. Defining a and b as chip evacuation force coefficients, and the chip evacuation torque model can be expressed as:

$$F_{ch}(z) = a(e^{bz} - 1) \tag{4.25}$$

where F_{ch} is the chip evacuation torque, z is the drilling depth, and the chip evacuation force coefficients a and b are positive. Considering drilling shear force F_c, the drilling torque can be expressed as:

$$F(z) = F_c + a(e^{bz} - 1) \tag{4.26}$$

By adopting this expression, the calibration test of different model parameters can be omitted, which facilitates iterative learning, and the accuracy of the model is guaranteed by continuously correcting coefficients in the real machining process. Using this model, based on the known maximal allowable chip evacuation torque, the maximum single drilling depth can be predicted.

4.4.2 *Chip Evacuation Process in Peck Drilling for Deep-Hole*

In practical machining, in order to avoid excessive chip evacuation force to cause tool broken, peck drilling is a frequently adopted method for deep-hole drilling to periodically decrease the chip evacuation force through multiple retracting and chip evacuation in order to get a deeper drilling depth. To establish the iterative learning method in real drilling cycles, it is necessary to study the changing law of chip evacuation force. Commonly, there are two types of peck drilling. The first one is peck drilling with equal depths which means setting a fixed single drilling depth Δd and retracting after entering this value. The other is peak drilling with variable depths which means using different depth settings Δd_i in each drilling step. This widely used strategy is to adopt the depth half of the previous step in each drilling step. Because the chip evacuation force increases with the increasing of drilling depth, this method can avoid drill breakage by successively decreasing the chip evacuation force as the drilling depth increases. The peak drilling cycle procedure is shown in Fig. 4.22.

In the drilling process, the workpiece material removed by cutting lips is entirely transformed into chips and pushed into the drill flute. The height of the chips that pack in the flute is denoted as the chip packing height. Suppose the increment of chip packing height dh is produced by the drilling depth dz, parameter $\lambda = dh/dz$ can be defined as the chip extrusion ratio. Since the drill occupies a portion of the space of the hole in the drilling process, the value of the chip extrusion ratio is larger than 1. In the one-step drilling process, as the chip extrusion ratio is greater than 1, a part of the chips is discharged out of the hole and does not interact with the hole wall. The chip packing height is then equal to the current drilling depth. Under this circumstance,

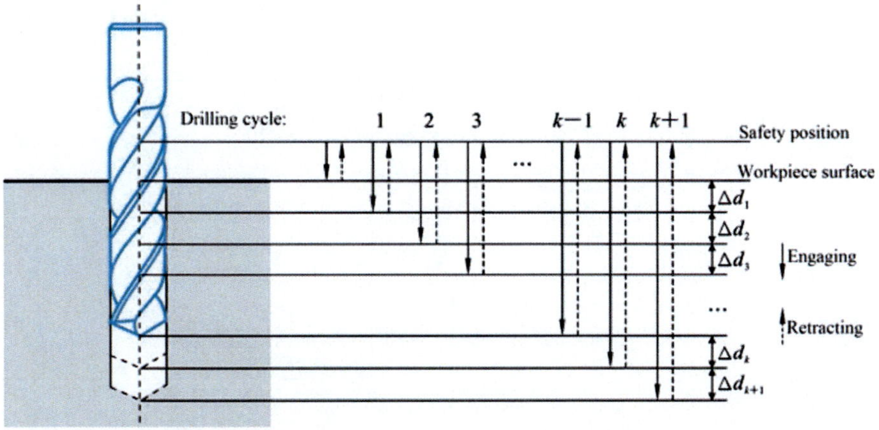

Fig. 4.22 Peak drilling cycle procedure

the chip evacuation force increases exponentially with the drilling depth according to the model law.

During the drilling cycle, since each drilling is carried out from the depth of the currently machined hole, the newly generated chips in the initial drilling stage are all limited to the space of the machined hole. At this time, the chip packing height is less than the current drilling depth, and the newly generated chips rub and squeeze with the hole wall and chip flute. Thus, the chip evacuation force increases with the chip packing height according to the model law. In addition, it is usually difficult to completely remove all chips in the chip flute during each retraction and chip evacuation process, resulting in a part of chips remaining in the chip flute during drilling. This part of the remaining chips hinders the flow of newly generated chips and increases the chip evacuation force. Then, the variation of the chip evacuation force with the drilling depth does not conform to the model law. The force generated by newly generated chips and remaining chips in the chip flutes after a certain drilling depth reaches the size of the chip evacuation force at the end of the last drilling. After that, the chip evacuation force will continue to increase exponentially on the basis of the previous value. The measured chip evacuation torque in a pecking drilling process is shown in Fig. 4.23. As there exists a certain period in the initial drilling process of each step where the chip evacuation torque increases from zero to the same value as the end of the previous drilling step, the whole drilling depth of peck drilling can be extended compared to the one-step drilling. This extended drilling depth in each drilling step is defined as the extended depth by chip removal (EDbCR), denoted as $\Delta z_{ex,k}$. The maximum total drilling depth of peck drilling can be described as the sum of the maximum one-step drilling depth and the EDbCR of each drilling step, which can be expressed as:

$$z_{\text{total}} = z_{\max} + \sum_{k=1}^{n-1} \Delta z_{ex,k} \tag{4.27}$$

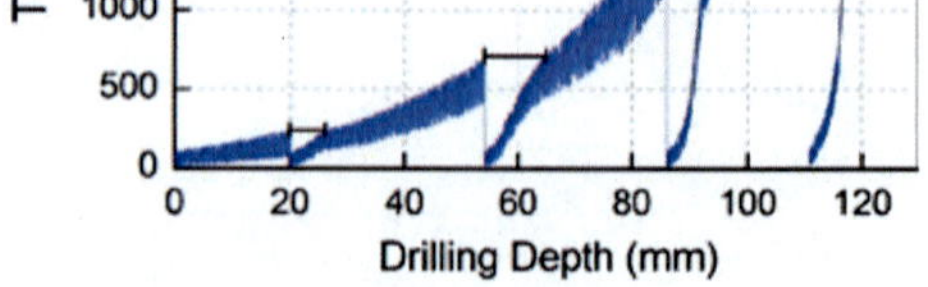

Fig. 4.23 Measured chip evacuation torque in peck drilling

where z_{total} is the maximum total drilling depth of peck drilling and $z_{\max}$ is the maximum one-step drilling depth.

In peck drilling, the chip removal process in each tool retracting operation removes a certain amount of chips, which causes the chip evacuation torque to decrease and go through a certain drilling depth to return to its original value. This is why pecking drilling is able to extend drilling depth compared with one-step drilling. In order to predict the peck drilling depth, the maximum one-step drilling depth and the EDbCR need to be known. In this book, the maximum one-step drilling depth is predicted by iterative learning on the chip evacuation torque model. When fitting the model by measured torque data, the torque data in the EDbCR are eliminated and only retained in accordance with the chip evacuation force model.

In the drilling cycle process, after finishing the single drilling, the chip packing height is regarded to be equal to the current drilling depth. Then the relation between the total height of all the residual chips and the EDbCR is:

$$z_{k-1} = h_{r,k} + \lambda \cdot z_{ex,k} \tag{4.28}$$

where z_{k-1} is the drilling depth of the previous drilling step and $h_{r,k}$ is the sum of the residual height of the chips in the flutes.

A parameter χ_k is introduced, which is defined as the proportion of the chips removed by the coolant in the previous step among the total chips packing in the flutes, which is denoted as the chip removal rate:

$$\chi_k = 1 - \frac{h_{r,k}}{z_{k-1}}, \chi_k \in (0, 1) \tag{4.29}$$

The chip removal rate reflects the performance of chip removal in the tool retracting process. Therefore, the closer to 1 the value of χ_k is, the better the chip removal effect will be. By solving Eqs. (4.28) and (4.29) simultaneously, the EDbCR can be derived as:

$$\Delta z_{ex,k} = \frac{\chi_k z_{k-1}}{\lambda} \tag{4.30}$$

Let $\alpha_k = \frac{\chi_k}{\lambda}$ be the extended depth coefficient by chip removal (EDCbCR), Eq. (4.30) can be expressed as:

$$\Delta z_{ex,k} = \alpha_k z_{k-1}, \alpha_k \in (0, 1) \tag{4.31}$$

Given the cutting fluid system, cutters, workpiece materials and cutting parameters, it can be considered that the chip removal rate and chip conversion ratio are only related to the drilling depth, so the EDCbCR under the above conditions is determined by the drilling depth. Then, by comparing the monitoring chip evacuation torque data of the two adjacent steps during the drilling cycle, the EDCbCR can be calculated according to the above formulas. The regression method can be used to obtain the law of the EDCbCR with the drilling depth as the iterative process carries on.

4.4.3 *Iterative Learning Method for Drilling Depth Optimization*

The goal of the iterative learning method is to obtain the maximum single drilling depth and the regression equation of the EDCbCR with the drilling depth, so as to combine the two to optimize the drilling depth value and avoid tool broken on the basis of the little number of retracting. The general process of drilling depth iterative learning and optimization is as follows. Firstly, setting a small initial drilling depth and calibrating the chip removal torque model coefficients using measured torque; Then, Predicting the maximum single drilling depth based on the calibrated model and giving the next drilling depth according to the iterative learning formula of drilling depth; In addition, combining the current and previous torque to recalibrate the model coefficients; What's more, predicting the maximum drilling depth and calculating the drilling depth of the next drilling; Lastly, according to the measured extended depth coefficients by chip removal of the previous drilling, predicting the EDbCR of the next drilling by the regression method. Repeating the above process, the model coefficients are continuously revised during the drilling cycle, and the prediction accuracy is gradually improved. As the drilling process carries on, the model is corrected by the measured data. In turn, the calibrated model is used to give the next drilling depth, forming a closed-loop of iterative learning.

During the iterative learning process, due to the lack of data in the previous iterative fitting model, the model's accuracy is not high and the maximum drilling depth predicted by the model may exceed the theoretical maximum drilling depth. The cutter may be broken if directly using the model to predict the maximum drilling depth. At the same time, if the drilling depth setting value is too large, even if the cutter does not break, it will still lead to excessive chips which could exceed the chip evacuation capacity of the machining system and resulting in chip removal difficulties and affecting the normal drilling. To avoid this problem and to ensure that each drilling depth is less than the theoretical maximum drilling depth, a modified Newton's method is proposed to optimize the drilling depth of each iteration. Newton's method is an important numerical method for solving nonlinear optimization problems. Its basic idea is to use the linear part to approximate the nonlinear function step by step. In order to obtain Newton's iterative formula of the drilling depth, the drilling torque model is expanded by Taylor at $z = z_k$, and the high-order terms are omitted. Then the torque model can be expressed as the following linear form:

$$F(z) = F(z_k) + F'(z_k)(z - z_k) \tag{4.32}$$

For deep-hole drilling, the goal of the optimization is to searching the maximum drilling depth z_{max} satisfying drilling torque. Therefore, the maximum allowable torque F_{max} is substituted into Eq. (4.32), then Newton's iterative formula of the drilling depth becomes:

$$z_{k+1} = z_k + \frac{F_{max} - F(z_k)}{F'(z_k)} \tag{4.33}$$

where $F'(z_k)$ is the gradient of the drilling torque. $F'(z) = abe^{bz}$ is a monotonically increasing function. In order to avoid cutter broken during the deep-hole drilling cycle, the gradient at the currently predicted maximum drilling depth is adopted as the searching direction. Because $F'(z_{max}) > F'(z)$, the gradient of the drilling torque at the maximum drilling depth is greater than that at all previous drilling depths. This method amplifies the growth rate of drilling torque during the iteration. In the first several iterations, the drilling depth is quite different from the target depth, and the accuracy of the fitted model is not high. The enlarged gradient ensures that the next drilling depth is less than the theoretical maximum drilling depth, which prevents the cutter broken due to excessive drilling depth during the drilling cycle. With the drilling cycle continuing, the drilling depth gradually approaches the maximum. The difference between the gradient $F'(z_{max})$ at the maximum drilling depth and the gradient $F'(z)$ at the current drilling depth gradually decreases. The error between the drilling depth predicted by using this gradient and the final theoretical maximum drilling depth is relatively small. Therefore, replacing $F'(z_k)$ in Eq. (4.33) with the gradient $F'(z_{max})$ at the maximum drilling depth, the iterative learning formula of the drilling depth is:

$$z_{k+1} = z_k + \frac{F_{max} - F(z_k)}{F'(z_{max})} \tag{4.34}$$

where $F(z_k)$ is the measured torque in the depth z_k, and z_{k+1} is the drilling depth at the next iteration. z_{max} can be calculated by the chip evacuation force model.

$$z_{max} = \frac{\ln\left(\frac{F_{max} - F_c}{a} + 1\right)}{b} \tag{4.35}$$

After each retracting and chip evacuation, it is necessary to carry out drilling after the EDbCR, so that the chip evacuation force can restore to the original value. Therefore, introducing the EDbCR in the iterative learning formula of the drilling depth, the final iterative learning formula can be written as:

$$z_{k+1} = z_k + \Delta z_{ex,k+1} + \frac{F_{max} - F(z_k)}{F'(z_{max})} \tag{4.36}$$

Substitute Eq. (4.31) into Eq. (4.36), then it becomes:

$$z_{k+1} = (\alpha_{k+1} + 1)z_k + \frac{F_{\max} - F(z_k)}{F'(z_{\max})} \tag{4.37}$$

In the first two drilling steps, the extended depth coefficients by chip removal can be set to zero, that is, $\alpha_1 = \alpha_2 = 0$, and α_{k+1} the ratio of the EDbCR in the $k + 1$th step to the drilling depth of the kth step. In the following steps, the EDCbCR can be obtained by regression analysis of its variation law in the previous steps. As the drilling process progresses, the chip evacuation force increases with the drilling depth. Thus, the incremental drilling depth in each step will gradually decreases when closing to the theoretical maximum drilling depth. The iterative learning process can be terminated when the difference value between the predicted maximum drilling depth and the current drilling depth is less than a set value $z_{\min}$, that is, $|z_{\max} - z_k| < z_{\min}$.

The flow of the deep-hole iterative learning method mentioned above is:

(1) Setting an initial drilling depth z_0;
(2) Drilling to the depth z_k and monitoring the drilling torque. Fitting the measured data according to the chip evacuation torque model and obtaining the model coefficients a and b;
(3) Predicting the maximum drilling depth $z_{\max}$ by the calibrated model in step (2), and calculating iterative gradient $F'(z_{\max})$;
(4) If $k = 0$ or $k = 1$, then $\alpha_k = 0$. Otherwise, operating regression analysis on $\alpha_{1,}\alpha_2, \ldots, \alpha_{k-1}$ to obtain the EDCbCR α_k;
(5) Substituting the iterative gradient $F'(z_{\max})$ in the step (3), the EDCbCR α_k and the maximum allowance torque into the drilling depth iterative learning formula to calculate the drilling depth z_{k+1} of the next iteration;
(6) Letting $k = k + 1$ and repeating the steps (2)~(5). Setting the minimum drilling depth $z_{\min}$ on one step. When satisfying the iterative terminal condition $|z_{\max} - z_k| < z_{\min}$, finishing the iterative learning process.

The above-mentioned deep-hole drilling iterative learning method is carried out during the drilling cycle of real hole processing. Compared to the model-based method, the proposed method is able to adjust the model coefficients according to the real drilling process so that the prediction accuracy is improved. Compared to the experimental data-based method, the proposed method can be carried out in the real peck drilling process, leaving out extra experimental works before the real drilling operation and reducing the cost. Compared with the online-monitoring-based tool broken protection method, it can learn from the monitoring data of the real machining process, and retains the advantages of data authenticity and real-time performance. At the same time, the drilling deep iterative learning formula can be used to provide drilling depth setting value before the next machining cycle to avoid cutter broken.

4.5 Process Optimization Method for Multi-hole Varying-Parameter Drilling

Nowadays, although various hole-making tools have been invented and applied to industry, about 75% hole making processes are still performed with traditional drilling method with twist drills. For drilling of the parts produced by difficult-to-cut materials with a large number of holes, called Ni-based superalloy such as turbine discs, compressor blisks and casings of an aeroengine, drill wear is a prominent problem to deal with. Thus, in the actual hole drilling process of key components of aeroengines, a large number of holes in these parts are drilled successively in one process to ensure their positional accuracy and machining efficiency. During the drilling of multi-hole parts, the tool wear and chip will naturally accumulate which could cause the tool wear and chip evacuation state to vary with the processing. At the same time, there are interactions between each hole and each drilling step. However, most of the drilling process parameter optimization methods at present are commonly focusing on the process of a single hole or single drilling step, if the fixed optimized parameters are adopted on all holes and drilling steps, the surface roughness of the holes is unstable and difficult to be satisfied and the tool load will fluctuate [6].

To address the abovementioned problems, a varying-parameter drilling process method (VPD) is proposed, which regards the multi-hole drilling process as a whole for global optimization. Variable process parameters that follow the hole surface quality and tool loads caused by tool wear and chip removal to adaptively adjust are adopted to achieve the purpose of ensuring processing quality, improving machining efficiency and reducing production costs.

4.5.1 Mathematical Model of Drilling Parameter Optimization

4.5.1.1 Solution of the Problem

In the VPD, the cutting parameters are varied for each hole to achieve a global optimum for the successive drilling process. Accordingly, the solution of the optimization problem is described as a sequence of spindle speed and feed rate. It can be expressed by the matrix below:

$$x = \begin{bmatrix} s_1 & s_2 & \cdots & s_n \\ f_1 & f_2 & \cdots & f_n \end{bmatrix}_{2\times n} \tag{4.38}$$

where x represents the solution of the optimization problem, n is the number of holes to be machined, s is spindle speed and f is feed rate. Each column of this matrix corresponds to the cutting parameters used for one hole. Compared to traditional fixed-parameter drilling, the cutting parameters s and f in the VPD are adjustable for different holes. The cutting parameters used for each hole can be flexibly varied to better fit the fast variation of drill wear in drilling of Ni-based superalloy, to get shorter machining time and better hole quality of all the holes.

4.5.1.2 Objective Function

To improve the manufacturing efficiency of a large number of holes, the optimization objective is to minimize the total processing time. In the successive drilling process, the processing time comprises three portions: machining time, idle motion time and tool changing time. The machining time is determined by the feed rate selected for each hole. The idle motion time consists of the time of the tool approaching, retracting and moving to the next holes. It is restricted by the maximum permissible feed rate of the machine tool and the length of the idle toolpath. Tool changing time refers to that when the drill gets severe worn, the drill has to be changed and it costs the time for tool presetting and adjustment. The total time for a tool change is determined by the frequency of tool failure that is related to the drill wear rate, and the average once tool changing time. It is a function of drilling depth, spindle speed and feed rate. Then, the objective function is given as:

$$t = \sum_{i=1}^{n} \frac{D}{s_i f_i} + T_c \cdot m + T_{idle} \tag{4.39}$$

where t is the total processing time of the successive drilling process, n is the total number of holes to be machined, i is the sequence number of each hole, D is the depth of the hole, s_i is the spindle speed for the i th hole, f_i is the feed rate for the i th hole, T_c is the tool changing time, m is the total number of tool change, T_{idle} is the idle motion time. For a specific multi-hole part, the path of the idle motion can be optimized as a traveling salesman problem (TSP) considering the distribution of the holes. There exists the shortest path for drilling all the holes. To reduce the total processing time, the shortest path and the maximum permissible idle feed rate can be adopted.

4.5.1.3 Constraints

As aforementioned, the surface quality of the holes is unstable due to the varied drill wear and needs to be controlled in the optimization. In workshops, the surface roughness is commonly adopted as the index of hole surface quality. As for the hole of a particular part, there is a tolerance requirement of the surface roughness. Therefore, the R_a value is the most widely used parameter of surface roughness, which is calculated as:

$$R_a = \frac{1}{L}\int_0^L |Y(l)|dl \tag{4.40}$$

where L is the sample length on the drilled hole wall, l is the ordinate along the sample length, Y is the ordinate of the profile curve along the sample length.

Besides, the constraints of spindle speed and feed rate that are related to cutting performance and machine tool property should also be considered in the practical situation. Then, the constraints of the optimization problem can be given by:

$$\begin{cases} R_a \le R_{a,\max} \\ s_{\min} \le s \le s_{\max} \\ f_{\min} \le f \le f_{\max} \end{cases} \tag{4.41}$$

where $R_{a,\max}$ is the tolerance of the hole surface roughness. The range of the cutting parameters $[s_{\min}, s_{\max}]$ and $[f_{\min}, f_{\max}]$ can be set as per the tool manufacturer's recommendation. As hole surface roughness is affected by various factors including cutting parameters, cutting force and tool wear, the ANN is used to develop the complex nonlinear relations between hole surface roughness and all these influence factors, which is detailed in the next section.

4.5.1.4 Fitness Function Construction

The optimization problem defined above is a constrained optimization problem. For the convenience of solving the problem, it is usually converted to an unconstrained problem. The penalty function is the most widely used constraint handling technique whereas it has the difficulty of balancing the optimization objective and the penalty to improve the convergence rate. When using the classic swarm intelligence algorithms like a genetic algorithm (GA) and particle swarm optimization (PSO) algorithm in the nonlinear constrained optimization problem, some candidate solutions may be infeasible and the proportion of the feasible solutions is varied as the searching progress which affects searching direction and convergence rate. In this case, there are two requirements for the penalty function: (a) infeasible solutions should move into the feasible zone rapidly, (b) feasible solutions should be restricted in the feasible zone and approach the optimum solution. Then, the procedure is detailed as follows.

(1) **Distance value**

To overall evaluate each particle's degree of deviation from the optimum solution, a new parameter denoted as the distance value is introduced. To this end, the objective value and constraint violation are normalized to balance their dimension. The objective value, that is, the total processing time t is normalized as follows:

$$\tilde{t}(x) = \frac{t(x) - t_{\min}}{t_{\max} - t_{\min}} \in [0, 1] \tag{4.42}$$

where x represents a solution in the searching algorithm, t is the value of the objective function, $\tilde{t}$ is the normalized objective value, $t_{\max}$ and $t_{\min}$ represent the possible maximum and minimum processing time, respectively.

The constraint violation of each candidate solution in this optimization is the sum of the deviation of surface roughness R_a from the tolerance $R_{a,max}$ for all the holes. It can be expressed as:

$$c(x) = \max\left\{0, \sum_{i=1}^{n} R_a^{(i)} - nR_{a,\max}\right\} \tag{4.43}$$

Then, it is normalized as:

$$\tilde{c}(x) = \frac{c(x)}{c_{\max}} \tag{4.44}$$

where $c_{\max} = \max\limits_{x} c(x)$ is the possible maximum value of constraint violation in real drilling. $\tilde{c}(x)$ is normalized constraint violation value which is confined within the range between 0 and 1.

Finally, the distance value d is formulated as:

$$d(x) = \begin{cases} \tilde{c}(x)\ if\ \alpha_f = 0 \\ \sqrt{\tilde{t}(x)^2 + \tilde{c}(x)^2}\ \text{otherwise} \end{cases} \tag{4.45}$$

where $\alpha_f = \frac{number\ of\ feasible\ particles}{swarm\ size}$ is the proportion of the feasible particles among all the candidate solutions. From Eq. (4.45), the distance value is increased when the value of the objective function and the constraint violation increase. If all the particles in the swarm are infeasible ($\alpha_f = 0$), the distance value is equal to the constraint violation.

(2) **Self-adaptive penalty**

In order to add a penalty to infeasible particles for a higher convergence rate, the self-adaptive penalty method (SAPM) is established, with the penalty function given as follow:

$$p(x) = (1 - \alpha_f)X(x) + \alpha_f Y(x) \tag{4.46}$$

where

$$X(x) = \begin{cases} 0\ if\ \alpha_f = 0 \\ \tilde{c}(x)\ \text{otherwise} \end{cases}$$

$$Y(x) = \begin{cases} 0 \text{ if x is a feasible individual} \\ \tilde{t}(x) \text{ otherwise} \end{cases}$$

From Eq. (4.46), two penalty values $X(x)$ and $Y(x)$ are added to make sure the penalty can be self-adaptive according to the proportion of feasible particles in the swarm: if there are a large proportion of feasible particles in the swarm, then the particles with high objective value are more penalized; and if there are few feasible particles, then those with high constraint violation are more penalized. The SAPM can help achieve two targets in the constrained optimization: (a) searching for more feasible solutions if there are few, (b) finding the optimum solution quickly if enough feasible particles are obtained.

(3) **Final fitness function**

The final fitness function $f(x)$ is obtained as the sum of the distance value $d(x)$ and the penalty function $p(x)$ as given below:

$$f(x) = d(x) + p(x) \tag{4.47}$$

From Eq. (4.45) to Eq. (4.47), when there are no feasible particles in the swarm ($\alpha_f = 0$), the fitness function becomes $f(x) = \tilde{c}(x)$. It only depends on the value of constraint violation. When the number of the feasible particles is increasing, the proportion of the objective value $\tilde{t}(x)$ in the fitness function is raised. The penalty on the particles is in-process adjusted for a fast convergence rate. In summary, the mathematical model of the optimization problem can be expressed by:

$$\begin{cases} find : x = \begin{bmatrix} s_1 & s_2 & \cdots & s_n \\ f_1 & f_2 & \cdots & f_n \end{bmatrix}_{2\times n} \\ min : f(x) = d(x) + p(x) \end{cases} \tag{4.48}$$

The optimization problem for the VPD involves different cutting parameters of all the holes. As the surface roughness of all the holes must be constrained within the tolerance, the boundary of the feasible zone is complex. Hence, the particles are prone to repeatedly moving into and out of the feasible zone in the searching process. The SAPM is able to adjust the fitness value according to the distance of the particles from the optimal solution and the proportion of feasible particles in the swarm. It is important for finding better solutions in the searching process so that the optimization can smoothly progress.

4.5.2 *Drilling Parameter Optimization Procedure*

After the fitness function of drilling parameter optimization is developed, the next issue is to find the optimum solution that minimizes the fitness function. A data-driven optimization method on the basis of the measured data from drilling experiments is presented to address the issue. The radial basis function (RBF) neural network used to predict the surface roughness can be trained with the dataset of drill wear, drilling forces and hole surface roughness at different cutting parameters. Then, the fitness value can be obtained by the proposed SAPM, and PSO can be conducted with its updating rule. The flow chart of the optimization procedure is illustrated in Fig. 4.24.

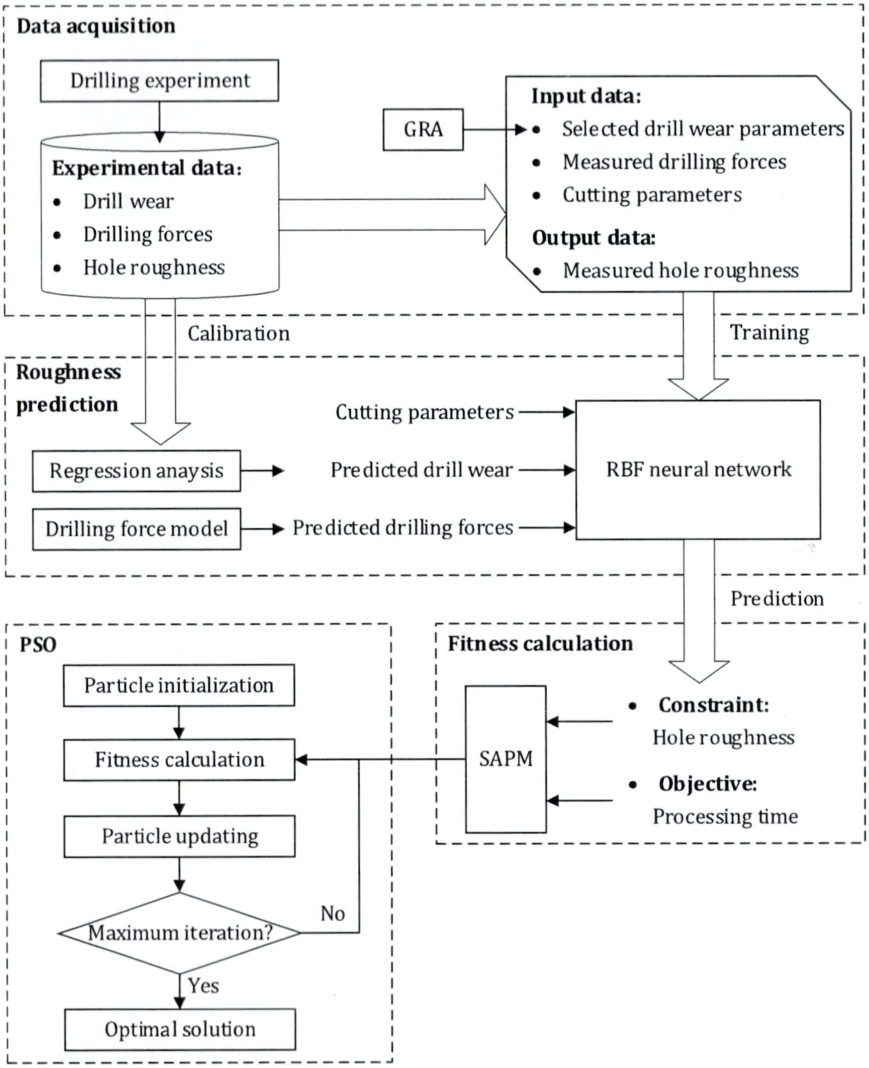

Fig. 4.24 Overall optimization procedure

4.5.2.1 Selection of Drill Wear Types

The geometry of a standard twist drill is illustrated in Fig. 4.25. Each type of drill wear will cause tool failure in the drilling process and may affect hole surface quality. Thus, all the wear types should be considered. The drill wear state can be quantitatively expressed by several parameters, which are called the drill wear parameters. They are the flank wear *VB*, the crater wear *KB*, the chisel wear C_ψ and V_ψ for two orthogonal directions, and the outer corner wear *W* (see in Fig. 4.26). Since there are two cutting lips for a standard twist drill, the mean value of the two drill wear parameters is used for analysis.

Before these drill wear parameters are used as input features of the RBF neural network, it is necessary to investigate their relevance with the hole surface roughness, because the input features with weak relevance will cause low prediction accuracy and extra computation cost. To this end, the relevance of these different drills wear types to hole surface roughness are analyzed with gray relation analysis (GRA) and

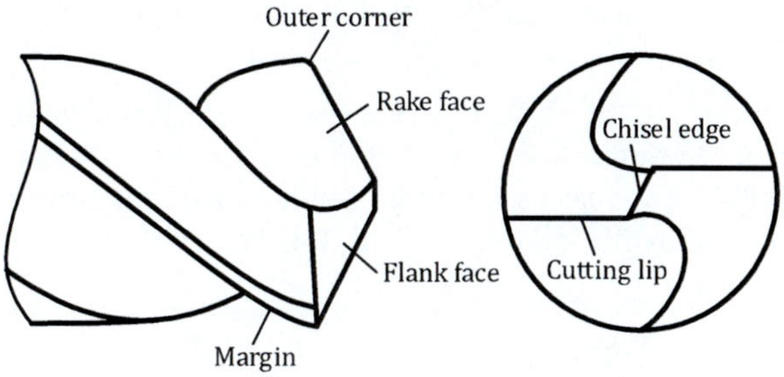

Fig. 4.25 Geometry of a standard twist drill

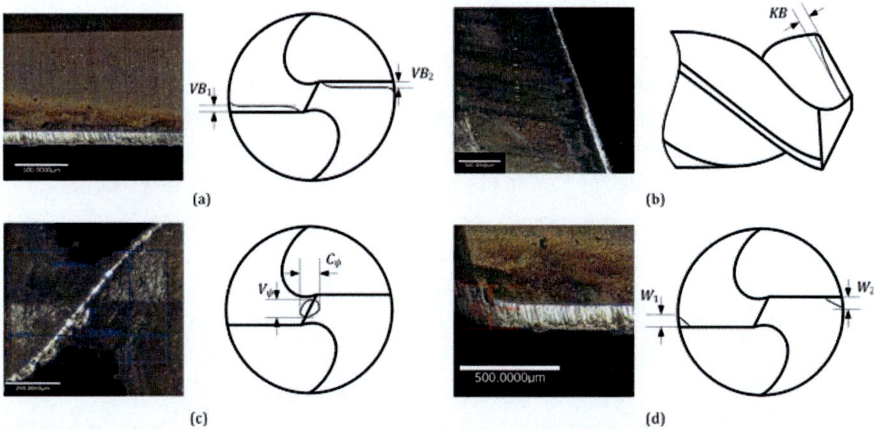

Fig. 4.26 Different drill wear types and their measurement: **a** flank wear, **b** crater wear, **c** chisel wear, **d** outer corner wear

the principal impact factors are selected as the input features. GRA is an analysis method in grey system theory that measures the uncertain correlations between one main factor (the reference sequence) and the other factors (the comparison sequences) in a given system. Since the relationship between various drill wear types and hole surface roughness are complex and uncertain, it can be considered as a grey system.

In the drilling parameter optimization problem, the reference sequence is the hole surface roughness, which is the variable to be investigated. The comparison sequences are the different types of drill wear, which are the influence factors. The reference sequence X_0 and each comparison sequence X_i can be expressed as follows:

$$X_0 = \left[R_a(1)\ R_a(2)\ \ldots\ R_a(n) \right]^T \tag{4.49}$$

$$\left[X_1\ X_2\ X_3\ X_4\ X_5 \right] = \begin{bmatrix} VB(1) & KB(1) & C_\psi(1) & V_\psi(1) & W(1) \\ VB(2) & KB(2) & C_\psi(2) & V_\psi(2) & W(2) \\ \vdots & \vdots & \vdots & \vdots & \vdots \\ VB(n) & KB(n) & C_\psi(n) & V_\psi(n) & W(n) \end{bmatrix} \tag{4.50}$$

where n is the number of the elements in the data sequence, and in this case referring to the number of measured holes. Since the unit and scale in each sequence are different, normalization is necessary to be conducted in order to keep the scale of data unified. The normalization formula is as follow:

$$X_i^*(k) = \frac{X_i(k) - \min\limits_k X_i(k)}{\max\limits_k X_i(k) - \min\limits_k X_i(k)} \tag{4.51}$$

where X_i^* denotes the normalized data, $X_0^*(k)$ and $X_i^*(k)$ are the normalized element in the reference sequence (hole surface roughness) and the comparison sequences (drill wear parameters), respectively,$i \in \{1, 2, 3, 4, 5\}$ and $k \in \{1, 2, 3, \ldots, n\}$. In GRA, the measure of the relevance between the reference sequence and the comparison sequence is evaluated as the grey relational grade. To calculate the grey relational grade, the grey relation coefficient $\xi_{0i}(k)$ is first introduced to measure the relevance between the corresponding elements in two sequences, and it can be given by:

$$\xi_{0i}(k) = \frac{\Delta_{\min} + \rho\Delta_{\max}}{\Delta_{0i}(k) + \rho\Delta_{\max}} \tag{4.52}$$

where $\Delta_{0i}(k) = \left| X_0^*(k) - X_i^*(k) \right|$ is the deviation of the element in the reference sequence and the comparison sequence, $\Delta_{\min} = \min\limits_i \min\limits_k \Delta_{0i}(k)$ and $\Delta_{\max} = \max\limits_i \max\limits_k \Delta_{0i}(k)$ are the minimum and maximum value of $\Delta_{0i}(k)$ in all the deviations, respectively. ρ is the distinguishing coefficient with the value between 0 and 1. The method has a larger distinguished ability when ρ is set as a less value. Generally speaking, $\rho = 0.5$. After the grey relational coefficient $\xi_{0i}(k)$ is obtained, the grey relational grade γ_{0i} can be calculated as the average value of $\xi_{0i}(k)$:

$$\gamma_{0i} = \frac{1}{n}\sum_{k=1}^{n}\xi_{0i}(k) \tag{4.53}$$

The grey relational grade γ_{0i} describes the degree of relevance between the reference sequence and a comparison sequence. From its definition, for each comparison sequence, the greater γ_{0i} is, the larger relevance it has with the reference sequence. The drill wear parameters with higher grey relational grade calculated from the measured data are selected as the input features of the neural network.

4.5.2.2 Prediction of Drill Wear and Drilling Forces

Drill wear and drilling forces are the input features of the RBF neural network for the prediction of the hole surface roughness. Drill wear is also involved in the calculation of tool changing time as mentioned in Sect. 4.5.1.2. Hence, drill wear and drilling forces should be calculated in the optimization. Since tool wear is a gradually cumulative process, the drill wear rate is introduced to calculate the drill wear parameters, which is defined as the incremental wear value of drilling per unit depth with the expression below:

$$r_i = \frac{dwear_i}{dl} \tag{4.54}$$

where r_i is the wear rate of each wear type, $wear_i$ denotes a particular drill wear parameter, l is drilling depth. The drill wear rate of different wear types can be estimated based on the measured drill wear data. As the wear rate r_i is time-varying, the drill wear can be calculated by integral:

$$wear_i = \int_0^l r_i dl \tag{4.55}$$

After the different drill wear parameters are obtained, the drilling forces can be predicted according to the model given below [7]:

$$\begin{cases} F = \delta_{a,lip}K_{a,lip}A + \delta_{\text{chisel}}K_{\text{chisel}}l_{chisel} \\ T = \left(\delta_{t,lip}K_{t,lip}A + \delta_{\text{corner}}K_{\text{corner}}l_{\text{corner}}\right)R \end{cases} \tag{4.56}$$

where F is the thrust force, T is the torque,A is the chip load, R is the drill radius, l_{chisel} and l_{corner} are the length of the chisel edge and the outer corner, $K_{a,lip}$ and $K_{t,lip}$ are the axial and tangential drilling force coefficients of the cutting lips, K_{chisel} and

K_{corner} are the drilling force coefficient of chisel edge and outer corners, $\delta_{a,lip}$, $\delta_{t,lip}$, δ_{chisel} and δ_{corner} are the four coefficients termed as the drill wear effect coefficients to describe the effects of various drill wear types on the corresponding drilling force coefficients. The drilling force coefficients and the drill wear effect coefficients can be calibrated in the drilling tests. When the drill wear is known, the drilling forces can be predicted as per this drilling force model. After then, the relations between drill wear, drilling forces and cutting parameters can be developed through experimental data.

4.5.2.3 Prediction of the Hole Surface Roughness with RBF Neural Network

The RBF neural network is a feedforward network with three layers: an input layer, a hidden layer and an output layer. The architecture of the network is illustrated in Fig. 4.27. The input parameters are spindle speed, feed rate, drilling thrust, drilling torque and the selected types of drill wear parameters. The input vector x_p is defined as:

$$x_p = \left[s \ f \ wear_1 \ wear_2 \ \ldots \ wear_k \ F \ T \right]^T \tag{4.57}$$

where $wear_1, wear_2, \ldots, wear_k$ represent the selected drill wear parameters that have large relevance with the surface roughness. Each node in the input layer receives the value from an input parameter, and the number of nodes is equal to the dimension

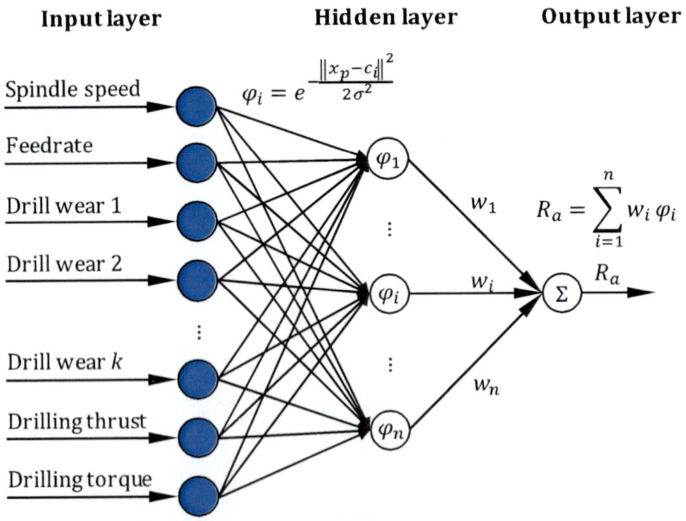

Fig. 4.27 Architecture of RBF neural network to predict hole surface roughness

of the input vectors. The output layer has only one node that is the hole surface roughness.

The hidden layer is responsible for the nonlinear processing of the input data. In the RBF neural network, the kernel trick is used to map the input data to a high-dimensional feature space so that the linear learning method can be applied for the nonlinear problem. Each input vector is transformed with the activation functions in the hidden layer, which in this case are the radial basis functions. The node in the hidden layer is called the radial node. The Gaussian kernel function is the most widely used radial basis function that can be expressed as:

$$\varphi_i = \exp\left(-\frac{\|x_p - c_i\|^2}{2\sigma^2}\right) \tag{4.58}$$

where x_p is the input vector, c_i is the hyper-center of the radial node in the hidden layer with the same dimension as x_p, $\|x_p - c_i\|$ represents the Euclidean distance between the input vector x_p and the center c_i, σ is called the spread parameter used to control the activation range of the hyper-centers, φ_i is the output value for each node in the hidden layer obtained from the transformed Euclidean distance. The hyper-centers are randomly chosen at the beginning and modified closest to the training sample during the training process. Due to the characteristic of the Gaussian kernel function, when the input vector x_p is close to the center c_i, the output value of the corresponding radial node is large and it means this center has a strong effect on the output node, and vice versa. Each radial node in the hidden layer is then connected to the output node with the weights w_i. The relation between the radial nodes and the output node (predicted surface roughness $\widehat{R}_a$) can be expressed as:

$$\widehat{R}_a = \sum_{i=1}^{n} w_i \varphi_i \tag{4.59}$$

These weights are updated during the training process to learn the relationship that exists between inputs and output. After the hyper-parameters of the network are obtained from the training process, the network can be used for the prediction of hole surface roughness with a specific input vector.

After the hole surface roughness is predicted with the RBF neural network, the fitness value expressed in Eq. (4.47) can be calculated with the SAPM. To be noted that, the network training is time-costing but conducted before the searching process with PSO, and the well-trained RBF neural network is then used for fitness calculation in the searching process.

4.5.2.4 Searching of Optimal Solution with PSO

The aforementioned SAPM can convert a constrained optimization problem into an unconstrained problem. While the hole surface roughness has a complicated nonlinear relationship with various influence factors including cutting parameters, drill wear and drilling forces, the Eq. (4.47) may be a non-convex function with multiple local optimums. In addition, since the hole surface roughness is calculated with an RBF neural network, it is difficult to formulate the fitness function given in Eq. (4.47). In this case, the classic gradient-based searching algorithm may be not effective. To this end, PSO is utilized to solve the optimization problem. PSO is a sort of meta-heuristic algorithm that has a strong global optimization capability proposed by Eberhart and Kennedy in 1995. Due to its good performance and concise expression, it has been widely used in many fields referring to optimization, design, control and data mining.

In PSO, the particles in the population represent potential solutions, and they move in the solution space of the problem to search for the optimal solution. In the searching process, each particle spreads its current location to other particles, and the location of the particle can be adjusted according to its current moving direction, self-experience and the information given by other particles. The population is gradually concentrated in the region containing high-quality solutions during the selection process.

Combining the above model and the PSO coupled with the self-adaptive penalty function, the process optimization problem for multi-hole varying-parameter drilling can be solved according to the following steps.

Step 1: Learning of machining data

The machining data can be utilized to select input features and train the RBF neural network of the hole surface roughness. Establishing the drilling force model and the relation between drill wear and cutting parameters, on this basis, the constraint violation of different particles can be calculated.

Step 2: Particle initialization

The solution of the optimization problem is expressed as a matrix in Eq. (4.38). The initial particles can be randomly generated with the set value of the spindle speed and feed rate within the feasible zone as per Eq. (4.41).

Step 3: Fitness calculation

The fitness value of each particle is obtained with the SAPM, in which the objective of processing time is calculated by Eq. (4.39) and the constraint of surface roughness is predicted with the RBF neural network. In each iteration of PSO, the optimum location of each particle and the global optimum location in the swarm are recorded.

Step 4: Particle updating

In PSO, the location of each particle is updated step by step according to the fitness value based on the iterative updating formula below:

$$v_{ij}^{k+1} = \omega v_{ij}^{k} + c_1 r_1 \left(pbest_{ij} - x_{ij}^{k}\right) + c_2 r_2 \left(gbest_{j} - x_{ij}^{k}\right) \tag{4.60}$$

$$x_{ij}^{k+1} = x_{ij}^{k} + v_{ij}^{k+1} \tag{4.61}$$

where i is the index of a particle, j is the index of the element in a particle, k is the current iteration number, x is the location of a particle (the candidate solution), v is the velocity of a particle, $pbest$ is the optimum location of a particle, $gbest$ is the global optimum location in the swarm, ω is the weight of the velocity, c_1 and c_2 are the learning factors, r_1 and r_2 are random numbers. The values of the parameters w, χ, c_1 and c_2 are set according to the performance of the algorithm in the searching process. At each iteration in this step, when the particles are updated by Eq. (4.59) and Eq. (4.60), the fitness values of new particles need to be calculated (go back to Step 3) and then the global optimum location $gbest$ and the optimum location of each particle $pbest$ are determined. If a certain value of spindle speed or feed rate in a particle is out of the range given in Eq. (4.41), it should be compulsively set as the corresponding boundary value of its constraint.

Step 5: Solution output and evaluation

In PSO, the termination condition of the iteration process in step 4 is the maximum iteration number. When the updating is executed at certain times, the global optimal solution is output as the final result. If the output sequence of cutting parameters is not satisfied the constraint in Eq. (4.41), there are three ways to deal with: (a) increase the maximum iteration number, (b) reinitialize the particles, (c) lose the constraint. To be noted that, as PSO is a meta-heuristic algorithm, it should have at least one feasible solution to guarantee the particles converging to a feasible solution. In the initialization step, when all the particles are infeasible, it is necessary to reinitialize the swarm until there exists a feasible particle.

4.6 Cyclic Iterative Optimization Method for Process Parameters

At present, the process parameter optimization methods of the milling process can be divided into offline optimization and online optimization. However, both the offline and online optimization methods have certain shortcomings and cannot be well applied in the actual machining.

The offline optimization method cannot adapt to the real part processing. It mostly bases on process prediction models, which require the prediction models to have high precision and wide applicability. In the real part machining process, factors such as the unevenness of the workpiece materials and the fluctuation of the process cannot be accurately predicted by models, simulations or experiments. Therefore, the application of the offline optimization method in actual processing is subject to certain restrictions.

The online optimization method cannot adapt to the global features of the part. It optimizes and regulates the process parameters based on the real-time monitoring data, which results in the method only being affected by local machining information. The optimization process lacks global consideration of the integral processing characteristics of the part, which makes the overall coherence of control poor. Moreover, the parameters that can be adjusted online during the milling process are limited, and it is impossible to fully optimize the control of the process. At the same time, in order to ensure the safety of the machining process, the online control process is generally conservative. In addition, the time lag requirement of online data analysis and adjustment is also one of the difficulties that restrict the application of the online optimization method.

In view of the characteristics of online and offline optimization, in order to give full play to the advantages of online and offline optimization, online solving and offline learning can be combined to perform iterative optimization of milling process parameters. Through online detection and identification of the instantaneous working conditions of the milling process, the feed rate optimization model can be solved by the actual working conditions identified. Then, feedback control is carried out on the milling process according to the solution results. After finishing the part, the process parameters in the part processing NC program are modified according to the recorded online solution results, and the solved process parameters are adopted as the initial parameters for the next part processing. An online solution process and an offline learning process based on the online solution data constitute a complete learning cycle, and the optimization of process parameters can be achieved through the iteration of the learning cycle [8].

The main process of online solution and control for milling processing feed rate is online monitoring the machining process through sensors and numerical control systems, and identifying instantaneous working conditions online by detecting processing; solving the feed rate optimization problem according to the current actual working condition obtained from the monitoring and recognition; performing real-time feedback control to the machining process based on the optimization results. This method does not need simulation, testing and trial cutting before the real part machining process, and all processes are completed online.

The main process of offline learning of milling process parameters is: synchronously matching the results of the online solution of the milling feed rate, the machining NC code and structural features of the part; calculating the maximum allowable spindle speed of each machined section according to the local structural features and tool performance constraints and setting the value as the new spindle speed for this section; recalculating the feed speed and correcting the spindle speed and feed speed on the NC code by the reselected spindle speed and the feed per tooth on the NC code obtained from the online solution; analyzing the corrected milling NC code and smoothing the process parameters to avoid severe acceleration and deceleration of the spindle speed and feed rate of the machine tool during processing, which may affect the machining quality and damage the cutter or machine tool. The whole procedure realizes learning and accumulation of process parameters, and provides a better initial NC code for the processing of the next part.

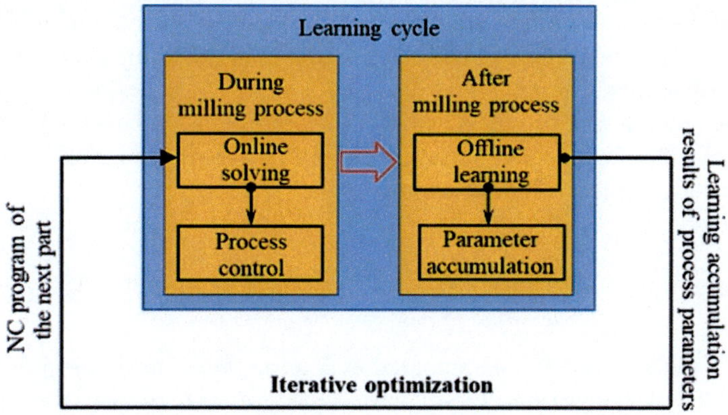

Fig. 4.28 The principle of learning cycle and iterative optimization for milling process parameters

The offline learning process corrects the NC codes based on the results of the online solution for the process parameters and the overall processing characteristics of the machined parts, so as to provide a better NC machining program for the next part processing. It not only ensures the accuracy of optimized process parameters but also overcomes the shortcomings of online solution and control.

The principle of iterative optimization of milling process parameters can be summarized as follows. Through a single learning cycle process, the milling process parameters are solved and learned and will be used for the next part of processing. Then the process parameters are constantly accumulated and updated based on a continuous iterative learning cycle, so as to realize the optimization of milling process parameters. The NC code of the part processing program is a carrier and concrete manifestation of part machining process knowledge. Therefore, the essence of iterative optimization of process parameters is also a process of learning and evolution of process knowledge. The principle of learning cycle and iterative optimization for milling process parameters is shown in Fig. 4.28.

4.6.1 Mathematical Model of Feed Rate Optimization

4.6.1.1 Objective Function for Optimization Problem

To optimize machining parameters in rough milling, machining efficiency is taken as the optimization objective. In the real part processing, total machining time for a single workpiece includes production preparation time, loading and unloading time, regulating time, cutting time and tool change time. Therefore, the machining time of a single workpiece can be expressed as:

$$T = T_p + T_L + T_a + T_m + T_C \tag{4.62}$$

where T is the total machining time; T_p is the production preparation time; T_L is the loading and unloading time; T_a is the regulating time; T_m is the cutting time; T_C is the tool change time.

The preparation time of a single workpiece refers to the process preparation time, which is obtained by equally spreading the process preparation time of a batch of parts to each workpiece. And the expression is:

$$T_p = \frac{T_s}{N_b} \tag{4.63}$$

where T_s is the total production preparation time, and N_b is the number of the parts.

The total tool change time of a single workpiece refers to the total time consumed by multiple tool changes in the machining, and its expression is:

$$T_C = T_c \cdot \left(\frac{T_m}{T_1} \right) \tag{4.64}$$

where T_c is the time consumed by a single tool change; T_m is the cutting time of a single workpiece; T_1 is the cutting time when the cutter becomes blunt.

In the machining time of a single workpiece, the preparation time is the process preparation time, which has nothing to do with processing parameters. The loading and unloading time and regulating the time of the part machining are the time consumed by the clamping and adjusting on the processing site and are independent of the processing parameters. The cutting time depends on the total toolpath length and feed speed, which means it is associated with the processing parameters. The total tool change time of a single part is related to the cutting time and tool blunting time, and it is also associated with the processing parameters.

The maximum production efficiency refers to the maximum number of parts per unit time, or the minimum time required to produce a single part. In the machining time of a single workpiece, only the cutting time and the total tool change time are related to the cutting parameters. The cutting time and tool blunting time of the total tool change time are also related to cutting parameters. In a certain range of feed rate, the influence of feed rate on the cutting time is much greater than that on the cutting time used for tool blunting. Therefore, the machining efficiency is directly affected by the cutting time. The minimum cutting time of a single part should be taken as the specific objective in the process of cutting parameter optimization.

In the milling process, when the feed rate is fixed, the cutting time of a single workpiece can be expressed as:

$$T_m = \frac{L_t}{f} \tag{4.65}$$

where L_t is the total toolpath length of a part machining, and f is the fixed feed rate.

When the feed rate varies with the processing, the cutting time of a single workpiece can be written as:

$$T_m = \int_0^{L_t} \frac{1}{f(L)} dL \tag{4.66}$$

where L is the length of the toolpath machined to any point of the part, and $f(L)$ is the feed rate when machining any point of the part.

Therefore, the objective function of the optimization problem can be expressed as:

$$\min T_m = \int_0^{L_t} \frac{1}{f(L)} dL \tag{4.67}$$

The toolpath which has been determined by the CAM process before machining cannot be changed, and the form and total length cannot be adjusted. Therefore, online milling parameter optimization and feedback control can only be achieved by controlling the feed rate in real-time, which means the optimal variable of the optimization problem is the feed rate.

4.6.1.2 Constraint Criteria for Optimization Problem

1. **Milling forces constraint**

In the machining process, the milling force varies with cutter rotation periodically. Therefore, in the research of machining process optimization, the maximum peak cutting force or mean cutting force is usually taken as the research object, and the milling force which is constant or less than the given value is regarded as the constraint. Due to the time-varying working conditions, the milling force changes with the development of the toolpath in the real part machining. Therefore, taking the resultant mean radial milling force on the cutter as the consideration standard, the resultant mean radial milling force on the cutter is limited with a given threshold value in the whole milling process. It is expressed in the form of constraint equation as:

$$g_1(f) = \left|\overline{F}(f)\right| - F_{lim} \leq 0 \tag{4.68}$$

2. **Tool wear rate constraint**

In the milling process, the cutting time T_1 used for tool blunting must be greater than a certain time limit, so as to avoid the reduction of production efficiency caused by frequent tool change, and also limit the increase of the tool cost. The tool life is related to cutting parameters, machined parts, tool materials and other working conditions, that is, the tool wear rate is related to the working conditions. In the real part machining process, the working conditions are changing. And the working condition at a certain moment determines the tool wear rate at that time, as well as affects the single service life of the tool. Therefore, we can only limit tool wear rate under current machining condition. In this way, tool life can be guaranteed within a single machining sequence. Tool wear rate can be expressed as:

$$r_{VB} = \frac{dVB}{dt} \tag{4.69}$$

$$g_2(f) = r_{VB}(f) - r_{VBlim} \le 0 \tag{4.70}$$

3. **Other constraints**

Except for milling forces constraint and tool wear rate constraint, the process coefficient optimization problem is also restrained by other factors.

(1) **Feed rate**

The feed rate in the processing is directly related to the property of the feed shaft of the machine tool and could also be affected by the maximum feed rate. Therefore, the feed rate needs to satisfy the limitation of the machining system which can be described as:

$$g_3(f) = f - f_{lim} \le 0 \tag{4.71}$$

where f_{lim} is the limit value for feed rate.

(2) **Applied force on machining system**

The milling force in the process can be restrained by both tool property and maximum milling force suffered by the machining system. From the point of view of machining system performance, the maximum peak value of milling force is limited to avoid damage to the machining system caused by excessive milling force.

$$g_4(f) = F_{\max}(f) - F_{mtlim} \le 0 \tag{4.72}$$

where $F_{\max}(f)$ is the maximum peak value of the milling force applied on the machining system, and F_{mtlim} is the maximum milling force that can be suffered by the machining system.

(3) **Milling torque**

The output torque of the machine tool spindle is directly related to the process parameters in the milling process, and the value is limited by the maximum spindle torque of the machine tool. Therefore, the spindle torque in the milling process needs to meet the requirements of the maximum spindle torque of the machine tool.

$$g_5(f) = M(f) - M_{lim} \leq 0 \tag{4.73}$$

where $M(f)$ is the instantaneous output torque of the spindle in the milling process, and M_{lim} is the limitation value of the machine tool's maximum spindle torque.

(4) **Milling power**

In the milling process, the output power of the machine tool is directly affected by the process parameters and is also limited by the maximum output power of the machine tool. Therefore, the output power in the milling process needs to meet the requirement of the maximum output power of the machine tool.

$$g_6(f) = P(f) - P_{lim} \leq 0 \tag{4.74}$$

where $P(f)$ is the instantaneous output power of the machine tool in the milling process, and P_{lim} is the limitation value of the machine tool's maximum output torque.

(5) **Cutting speed**

The cutting speed is directly related to the spindle speed and tool diameter in the milling process, and its value is limited by the maximum spindle speed of the machine tool. Therefore, the cutting speed needs to meet the requirement of the machining system on cutting speed.

$$g_7(f) = \upsilon_c(f) - \upsilon_{clim} \leq 0 \tag{4.75}$$

where $\upsilon_c(f)$ is the instantaneous cutting speed, and υ_{clim} is the limitation value of machining system on cutting speed.

4.6.1.3 Mathematical Model for Optimization Problem

Based on the obtained objective function and constraint conditions of the optimization problem, the mathematical model for the optimization problem can be described as:

$$\begin{cases} \min T_m = \int_0^{L_t} \frac{1}{f(L)} dL \\ \text{s.t.} g_i(f) \le 0, i = 1, 2, \ldots 7 \end{cases} \tag{4.76}$$

4.6.2 *Online Solving for Feed Speed Optimization Problem*

The offline optimization method cannot adapt to the real machining process. The online optimization solution process can obtain the real working conditions, based on which the process parameters can be optimized and controlled. Thus, the difficulties that offline optimization cannot accurately predict various factors could be solved.

4.6.2.1 Online Solving Process for Optimization Problem

In the milling process, the online solution of feed rate optimization is essential to determine the milling feed rate of each point on the toolpath in real time. The online solution of feed rate optimization is the process of determining the next feed rate according to the current working condition information and the output response of the machining system. In the process of solving, in order to ensure the shortest machining time, it is necessary to make the milling feed rate of each point on the toolpath to get the maximum value. However, the value of the milling feed rate must satisfy all the constraints.

In the constraint criteria analyzed before, not all constraints can play a role in the optimization process. A series of constraints $g_i(f) \le 0, i = 1, 2, \ldots 7$ are redundant, and some of which are too broad and can be satisfied naturally. As the rigidity and processing capacity of machining systems such as machine tools and fixtures are generally far beyond the range that the workpiece-tool subsystem can bear in the machining process, that is, the weak link that bears the machining impact is mainly concentrated on the workpiece-tool subsystem. Therefore, in the process of solving the parameter optimization problem, the constraints such as feed speed, applied force on machining system, milling torque, milling power and cutting speed can be abandoned, and only the constraints related to the workpiece-tool subsystem, namely milling force and tool wear rate constraint, can be considered.

The essence of online solution and control for milling parameters optimization is a control theory problem, that is, calculating the regulation of process parameters according to the optimal goal and limitation criterion of the machining process. The online solution procedures of parameter optimization are as follows: (1) Calculating the adjustment value of feed rate according to the monitored average milling force, the limitation value of the milling force borne by the cutter and the influence of time-varying working conditions on the milling force; (2) Adding the calculated feed rate adjustment value to the current feed rate to obtain a new one; (3) Calculating the new tool wear rate according to the influence of time-varying working conditions on the

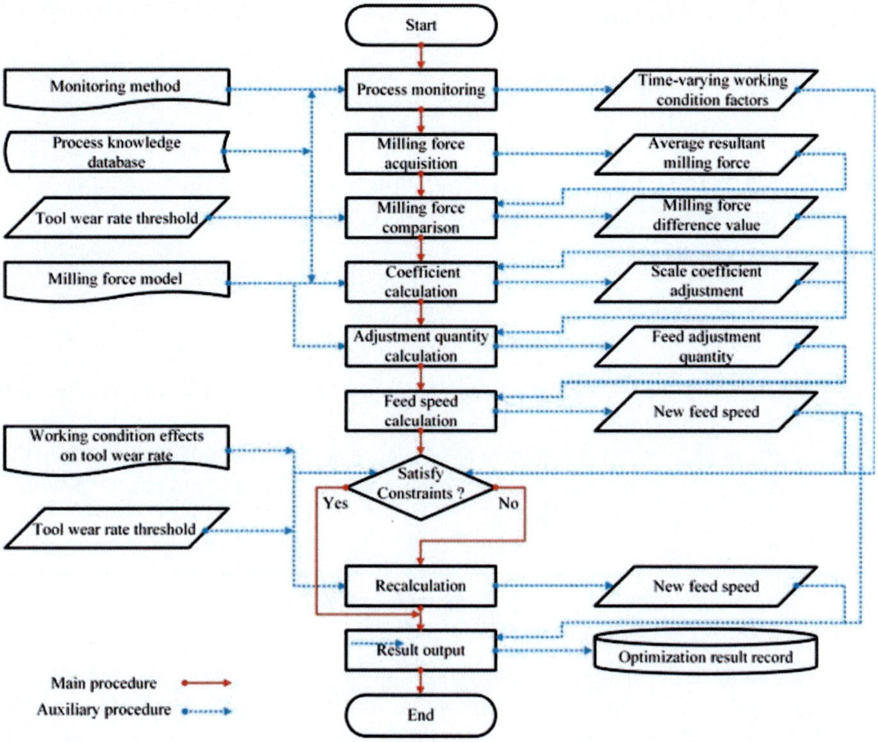

Fig. 4.29 The online solving procedures of milling feed rate optimization and the data flow between each step

tool wear rate in the milling process, and verifying whether the optimized feed rate meets the constraint criterion of tool wear rate; (4) If it is satisfied, the optimized feed rate will be taken to the next moment. (5) If it is not satisfied, the new feed rate will be recalculated according to the constraint criterion of tool wear rate, and the value will be taken as the feed speed at the next moment.

The online solving procedures of milling feed rate optimization and the data flow between each step are shown in Fig. 4.29. The feed rate optimization problem is solved online to obtain the optimal milling feed rate, and then the milling process is regulated by the feedback control module.

The above flow is a single online optimization solution and real-time feedback control process. In the real milling process, the feed rate will be continuously solved and feedback-controlled online until the end of machining.

4.6.2.2 Solving Method

1. **Milling force acquisition**

In the real milling process, the milling forces in X and Y directions could generally be monitored by force sensors. When the rotating angle of the milling cutter is φ, the measured milling force can be expressed in vector form as:

$$\boldsymbol{F}_{\text{meas}}(\varphi) = \begin{bmatrix} F_{x,\text{meas}}(\varphi) \\ F_{y,\text{meas}}(\varphi) \end{bmatrix} \tag{4.77}$$

where $\boldsymbol{F}_{\text{meas}}(\varphi)$ is the measured milling force vector in the OXY plane when the rotating angle of the cutter is φ.

The average milling force vector in OXY plane can be obtained by calculating the average value of the measured milling force vector when the cutter rotates within one round, and it can be expressed as:

$$\overline{\boldsymbol{F}_{\text{meas}}} = \frac{1}{2\pi}\int_0^{2\pi} \boldsymbol{F}_{\text{meas}}(\varphi)d\varphi \tag{4.78}$$

The resultant force is the modulus length of the force vector. Then, the average resultant force of the milling cutter measured during the machining process is the module length of the measured average milling force vector in the OXY plane.

$$\left|\overline{\boldsymbol{F}_{\text{meas}}}\right| = \sqrt{\overline{\boldsymbol{F}_{\text{meas}}}^T \cdot \overline{\boldsymbol{F}_{\text{meas}}}} \tag{4.79}$$

2. **Relationship between average milling resultant force and feed speed**

In the milling process, the average milling force vector can be written as the linear form of the feed per tooth, but there is no linear relationship between the resultant of the average milling force and the feed per tooth. Therefore, after conducting local linearization to the functional relationship between the resultant of average milling force and the feed per tooth, namely the first-order Taylor expansion, the relationship can be expressed as:

$$\left|\overline{\boldsymbol{F}}\right|(f_z) = \left|\overline{\boldsymbol{F}}\right|(f_{z,0}) + \frac{\partial\left|\overline{\boldsymbol{F}}\right|(f_{z,0})}{\partial f_z} \cdot (f_z - f_{z,0}) + R_1(f_z) \tag{4.80}$$

where f_z is the feed per tooth, $f_{z,0}$ is the feed per tooth at a certain point in the milling process, $\left|\overline{\boldsymbol{F}}\right|(f_z)$ is the functional relationship between the resultant of average milling force and feed per tooth, $\left|\overline{\boldsymbol{F}}\right|(f_{z,0})$ is the resultant of average milling force at a certain point in the milling process, $R_1(f_z)$ is the remainder of the first-order Taylor expansion.

The first partial derivative of the resultant of average milling force to the feed per tooth is:

$$\frac{\partial\left|\overline{\boldsymbol{F}}\right|}{\partial f_z}=\frac{1}{2\left|\overline{\boldsymbol{F}}\right|}\cdot\frac{\partial(\overline{\boldsymbol{F}}^T\cdot\overline{\boldsymbol{F}})}{\partial f_z} \tag{4.81}$$

Combined with the milling force formula, the first order Taylor expansion of the functional relationship between the resultant of average milling force and feed per tooth can be described as:

$$\left|\overline{\boldsymbol{F}}\right|(f_z)=\left|\overline{\boldsymbol{F}}\right|(f_{z,0})+\frac{\overline{\boldsymbol{F}}^T(f_{z,0})\cdot\boldsymbol{G}_c\cdot\boldsymbol{K}}{\left|\overline{\boldsymbol{F}}\right|(f_{z,0})}\cdot(f_z-f_{z,0})+R_1(f_z) \tag{4.82}$$

where $\boldsymbol{G}_c$ is the geometric influence matrix of the average shear force, $\boldsymbol{K}$ is the shear force coefficient matrix.

3. **Calculation for optimal feed speed**

According to Eq. (4.82), the resultant of average milling force and the feed per tooth have an approximately linear relationship locally, that is, the increment of the resultant of average milling force is approximately proportional to the increment of the feed per tooth.

$$k=\frac{\overline{\boldsymbol{F}}^T(f_{z,0})\cdot\boldsymbol{G}_c\cdot\boldsymbol{K}}{\left|\overline{\boldsymbol{F}}\right|(f_{z,0})} \tag{4.83}$$

When calculating the optimized feed per tooth, this proportional relationship can be used to calculate the adjustment value of the feed per tooth through the difference between the actual measured resultant of radial average milling force and its limitation value, that is

$$\delta f_z=(F_{\lim}-\left|\overline{\boldsymbol{F}}_{meas}\right|)k_p \tag{4.84}$$

where δf_z is the adjustment value of feed per tooth, $F_{\lim}$ is the limitation value of the radial force borne by the cutter, $\left|\overline{\boldsymbol{F}}_{meas}\right|$ is the resultant of radial average milling force which is obtained by online monitoring and real-time calculating, k_p is the proportional coefficient of the adjustment value for feed per tooth and it is also the reciprocal of k.

In the online solving procedures of milling feed rate optimization (seen in Fig. 4.29), the compassion of cutting force is calculating the difference between the radial average milling force and its limitation value, that is $F_{lim}-\left|\overline{\boldsymbol{F}}_{meas}\right|$.

Based on the calculated radial average milling force $\overline{\boldsymbol{F}}_{meas}$, the proportional coefficient of the adjustment value for feed per tooth k_p can be expressed as:

$$k_p = \frac{\left|\overline{\boldsymbol{F}_{meas}}\right|}{\overline{\boldsymbol{F}_{meas}}^T \cdot \boldsymbol{G}_c \cdot \boldsymbol{K}} \tag{4.85}$$

The proportional coefficient of the adjustment value for feed per tooth k_p is not only related to radial average milling force $\left|\overline{\boldsymbol{F}_{meas}}\right|$, but also related to the geometric influence matrix of the average shear force $\boldsymbol{G}_c$ and the shear force coefficient matrix $\boldsymbol{K}$. $\boldsymbol{G}_c$ is associated with constant working conditions, such as cutter parameters and milling engagement forms, and is also affected by time-varying working conditions, like process parameters. $\boldsymbol{K}$ is related to the constant working conditions, such as tool coating and workpiece materials.

When calculating k_p, $\overline{\boldsymbol{F}_{meas}}$ and $\left|\overline{\boldsymbol{F}_{meas}}\right|$ can be obtained by online monitoring and real-time calculating (seen the acquisition of milling force in Fig. 4.29). $\boldsymbol{K}$ and $\boldsymbol{G}_c$ need to be matched in the process knowledge base according to the steady working conditions of the current actual machining process. The determination of $\boldsymbol{G}_c$ also need to consider the time-varying working condition of the current machining state, and the time-varying working condition can be obtained by detecting machining (seen the process detecting in Fig. 4.29).

After obtaining $F_{lim} - \left|\overline{\boldsymbol{F}_{meas}}\right|$ and k_p, the adjustment value of feed per tooth can be calculated by Eq. (4.84), as well as the optimized new milling feed rate.

The optimized new feed per tooth can be obtained by superimposing the adjustment value of the feed per tooth on the current feed per tooth:

$$f_z^{opt} = f_z^{crt} + \delta f_z \tag{4.86}$$

where f_z^{opt} is the optimized new feed per tooth, f_z^{crt} is the current actual feed per tooth.

The current actual feed per tooth can be calculated by the current actual milling feed rate and spindle speed:

$$f_z^{crt} = \frac{f^{crt}}{n^{crt} N} \tag{4.87}$$

where f^{crt} is the current actual milling feed rate, n^{crt} is the current actual spindle speed, N is the number of teeth of the cutter which is fixed and has been determined before machining. Substitute Eq. (4.87) into Eq. (4.86), then the optimized new feed per tooth can be presented.

The optimized new feed rate can be calculated by the optimized new feed per tooth and spindle speed:

$$f^{opt} = f^{crt} + \delta f_z N n^{crt} \tag{4.88}$$

The current actual milling feed rate and spindle speed are time-varying conditions that can be obtained by process detecting in Fig. 4.29.

4. **Verification for tool wear rate constraint criteria**

In the milling feed rate optimization problem, in addition to the constraint criterion of the radial average milling force for the cutter, there is another constraint, namely the tool wear rate. After calculating the optimized new milling feed rate according to the first constraint criterion, it needs to be verified based on the second constraint criterion.

Tool wear rate is defined as the increment of tool flank wear per unit cutting time, that is, the derivative of tool flank wear to machining time.

$$r_{VB} = \frac{dVB}{dt} \tag{4.89}$$

where r_{VB} is the tool wear rate, *VB* is the tool flank wear of the cutter, *t* is the cutting time.

Tool wear rate can be described as the function of time-varying working conditions.

$$r_{VB} = r_{VB}(f_z, v_c, VB) \tag{4.90}$$

where f_z is the feed per tooth, v_c is the cutting speed.

Substituting the optimized new feed per tooth, current cutting speed, and tool flank wear into Eq. (4.90), the corresponding tool wear rate can be calculated. The optimized new feed per tooth can be computed by the optimized new milling feed rate and the current spindle speed. The current cutting speed can be calculated according to the current milling spindle speed and cutter radius.

The tool wear rate corresponding to the optimized new milling feed rate is calculated by Eq. (4.90), which is substituted into Eq. (4.70) to verify whether the tool wear rate meets the constraint. If the constraint of Eq. (4.70) is proper, the wear rate can be taken as the result of optimization. If the constraint of Eq. (4.70) is improper, the wear rate does not satisfy the constraint. Then, it is necessary to recalculate the optimized results of the milling feed rate according to the tool wear rate constraint criterion.

5. **Re-calculation for milling feed speed**

If the optimized new milling feed rate cannot meet the tool wear rate constraint criterion, it is necessary to recalculate the optimization result of the milling feed rate according to the tool wear rate constraint criterion, that is, to find the maximum milling feed rate that can meet the tool wear rate constraint.

The equation corresponding to the constraint of tool wear rate is solved by using the single point chord secant method to calculate the milling feed rate satisfying the conditions. Taking the current feed rate f_z^{crt} in the milling process as the initial point of the single point chord secant method, that is, the fixed iterative point, and taking the optimized feed rate according to the milling force constraint criterion as the first iterative point of the single point chord secant method, it can be expressed as follows:

$$\begin{cases} f_z^0 = f_z^{crt} \\ f_z^1 = f_z^{opt} \end{cases} \tag{4.91}$$

In the single point chord secant method, the process of iteratively calculating the $j + 1$th solution through the jth solution is to replace the original equation expression with the chord secant line connecting the initial fixed point and the corresponding point of the jth solution to perform a linear solving and take the root of the straight line as the $j + 1$th solution of the iterative method. The iterative formula of the solution is:

$$f_z^{j+1} = f_z^j - \frac{[r_{VB}(f_z^j) - r_{VB}](f_z^j - f_z^0)}{r_{VB}(f_z^j) - r_{VB}(f_z^0)} \tag{4.92}$$

where f_z^j is the jth solution of the single point chord secant method, f_z^{j+1} is the $j + 1$th solution of the single point chord secant method.

When judging the stop of the iterative solution, in addition to the accuracy of the control result, it is also necessary to ensure that the feed per tooth meets the tool wear rate constraint. After obtaining the feed per tooth, the corresponding milling feed rate can be calculated based on the current actual milling spindle speed.

The function between the tool wear rate and feed per tooth is an increasing function, and the recalculated milling feed rate is less than the milling feed rate computed according to the milling force constraint. Therefore, there is no need to verify the constraint criterion of the average milling force, which means the result of the recalculation is the final one.

According to the result of verifying the tool wear rate constraint criterion, the feed rate optimization result can be determined. The optimization result will be selected from the feed rate calculated from the milling force and the tool wear rate constraint criterion. After confirming the optimized final milling feed rate, it is output from the optimization solution process to the feed rate feedback control module, and at the same time, it is recorded in the storage medium to provide a data basis for offline learning of milling feed rate.

4.6.3 Offline Learning and Iterative Optimization for Process Parameters

The offline learning of process parameters refers to the process parameters in the NC code are corrected according to the process parameters obtained by online solving in the offline environment after the part machining is completed, so as to realize the learning and accumulation of the process parameters and provide better initial NC code for the next part processing.

The main procedure of offline learning for milling process parameters is as follows:

(1) Optimization matching. Match the online solution result of milling feed rate with the NC code of the part according to the code line, that is, obtain the optimization result corresponding to each line of NC code. Then match the NC code with the structure of the part to get the part machining position corresponding to each line of code. Finally, acquire the synchronous matching relationship between the parameter solution result, NC code and the part structure to provide support for the spindle speed selection and parameter modification.
(2) Spindle speed selection. Perform chatter stability simulation on the machined part to determine the maximum allowable speed of each position of the workpiece. Confirm the reselected spindle speed for each line of NC code according to the matching relationship between the NC code and the part structural. Select larger spindle speed to improve machining efficiency on the premise of satisfying the constraints of allowable linear speed, tool wear rate and maximum spindle speed.
(3) Parameter modification. Capture the optimized feed per tooth of each point on the NC code according to the online solution result and the matching relationship with the NC code, and calculate the new feed rate combining the reselected spindle speed. Correct the spindle speed and feed rate in each line of NC code based on the reselected spindle speed and recalculated feed rate.
(4) Parameter smoothing. Conduct smoothing for the variation of process parameters to avoid the excessive change of the spindle speed and feed rate of the machine tool during the machining process, which affects the machining quality or even damages the cutter or machine tool. Analyze process parameters along the revised NC code and execute uniform smoothing transition to the change of process parameters to improve the stability of the movement of the machine tool and reduce drastic mutations.

The above process realizes learning and accumulation of process parameters and provides better original NC code for the next part. The main procedures of offline learning for milling process parameters are shown in Fig. 4.30.

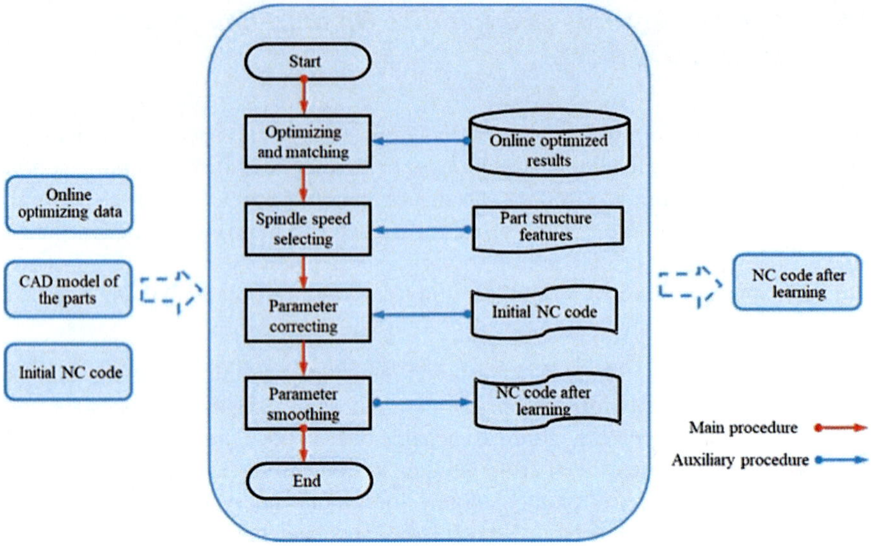

Fig. 4.30 The flow of offline learning for feed rate

References

1. ZHOU X. Research on Dynamic Response Prediction and Control in the Milling Process of Ring-shaped [D]. Xi'an; Northwestern Polytechnical University, 2017.
2. HOU Y H, ZHANG D H, MEI J W, et al. Geometric modeling of thin-walled blade based on compensation method of machining error and design intent [J]. Journal of Manufacturing Processes, 2019, 44: 327–336.
3. HOU Y H, ZHANG D H, ZHANG Y. Error compensation modeling amd learning control method for thin-walled part milling process [J]. Journal of Mechanical Engineering, 2018, 54(17): 108–115.
4. HAN C, ZHANG D H, LUO M, et al. Chip evacuation force modelling for deep hole drilling with twist drills [J]. The International Journal of Advanced Manufacturing Technology, 2018, 98(9–12): 3091–3103.
5. HAN C, LUO M, ZHANG D H, et al. Iterative learning method for drilling depth optimization in peck deep-hole drilling [J]. Journal of Manufacturing Science and Engineering, 2018, 140(12): 1–12.
6. HAN C. Condition Monitoring and Learning Optimization of Drilling Process for Aeroengine Key Components [D]. Xi'an; Northwestern Polytechnical University, 2020.
7. HAN C, LUO M, ZHANG D H, WU B H. Mechanistic modelling of worn drill cutting forces with drill wear effect coefficients [C], Procedia CIRP, 2019, 82: 2–7.
8. HOU Y F. Detection and monitoring recognition and learning optimization method of time-varying factors in milling condition [D]. Xi'an; Northwestern Polytechnical University, 2015.

Chapter 5
Dynamic Response Prediction and Control for Machining Process

The dynamic response generally refers to the continuous output signal from the initial state to the final state generated by the system under the action of input excitation. As for the milling process of thin-walled parts, the dynamic response of the process system is the vibration of the process system excited by the intermittent impact of the cutter teeth or the fluctuation of chip thickness, among which chatter and forced vibration are the most important. In the metal cutting process, even if there is no periodic external force, due to the interaction of excitation and feedback, intense vibration often occurs between the workpiece and the tool which is called milling chatter. The amplitude and frequency of the chatter are determined by the process system itself. In addition, the cutter teeth continuously excite the workpiece during the periodical entering and exiting of low rigidity workpiece, thereby generating vibration, that is, forced vibration. The frequency and amplitude of forced vibration are related to periodic cutting force, machining system stiffness, damping and frequency. These two kinds of vibration exist in the low rigidity machining system together, and both have an adverse effect on the surface roughness and machining accuracy, and damage the cutters and machine tools and reduce production efficiency. For a long time, many scholars have devoted themselves to the research of machining vibration theory, in order to reveal the mechanism and law of its generation and control it [1].

5.1 Control Method of Dynamic Response for Machining Process

It can be seen that, from the basic dynamic Eq. (2.1), we can adjust ***M, C, K, F*** to achieve the ideal dynamic response state of the machining system. Among them, the cutting force ***F*** can be regulated by the cutting parameters and the excitation control of the teeth. Specifically speaking, it includes the optimization of cutting parameters and the control of cutter pitch angle. And the control of ***M, C,*** and ***K*** can be achieved

D. Zhang et al., *Intelligent Machining of Complex Aviation Components*, Research on Intelligent Manufacturing, https://doi.org/10.1007/978-981-16-1586-3_5

Table. 5.1 Dynamic response of the process system and its control method

Parameters	Control methods
F	Parameter optimization, tool structure optimization (pitch angle)
M	Additional masses
C	Auxiliary support fixture, additional damping support
K	Auxiliary support fixture, additional damping support

by adjusting the clamping and changing the process system attributes. The dynamic response of the process system and its control method are shown in Table 5.1.

5.2 Alternating Excitation Force During Milling

In the semi-finishing and finishing process of aero thin-walled parts, due to the small machining allowance, the cutter teeth will intermittently enter and exit from the workpiece to generate alternating excitation force and cause machining vibrations.

5.2.1 Alternating Excitation Force

In the process of semi-finishing and finishing of complex structural parts of aero-engines, intensive machining toolpaths are usually planned to ensure that the residual height meets the design requirements. At this time, the tool will always be in a smaller radial and axial cutting depth, as shown in Fig. 5.1. In this case, the cutting state of

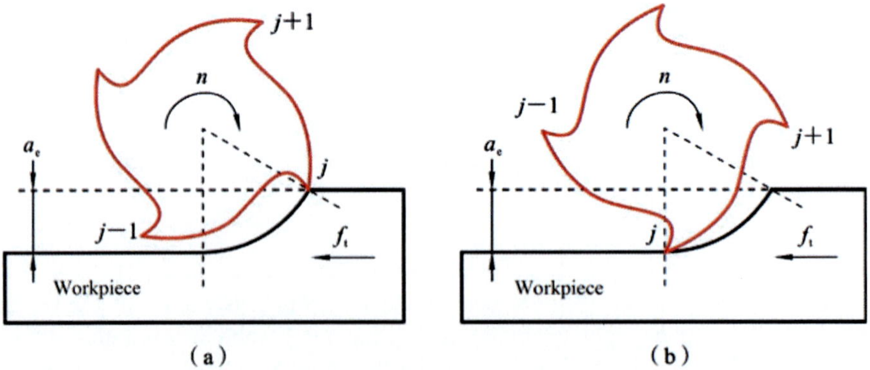

Fig. 5.1 Tooth entering and exiting from the workpiece **a** tool entering **b** tool exiting

the tool is intermittent cutting, that is, when the tooth j starts to enter the material, the tooth $j-1$ has already exited from the material, as shown in Fig. 5.1a. When the tooth j exits from the material, the tooth $j+1$ has not yet started to cut the material, as shown in Fig. 5.1b.

When the tooth j engages in the cutting stage, the chip thickness, as well as the cutting force corresponding to the disk element, is not zero. And when the tooth j finishes cutting and the tooth $j+1$ has not yet started cutting, the chip thickness and the cutting force corresponding to the disk element is equal to zero. Therefore, in the milling process, the rotation of the cutter causes each tooth to continuously cut in and out the workpiece. This highly intermittent cutting brings continuous impact excitation to the workpiece and causes the cutting force to fluctuate, which leads to the inevitable forced vibration during the machining process.

Noting that n is the spindle speed, and N is the number of tool teeth, then the period of cutting in and out the workpiece can be represented by the tooth passing frequency f_{tpf}:

$$f_{tpf}=\frac{nN}{60} \tag{5.1}$$

5.2.2 *Characterization and Decomposition of Alternating Excitation Force*

According to the cutting force mechanical model from Altintas [2], the radial, tangential and axial cutting forces acting on the cutting-edge element can be expressed as a function of machining time t and axial height z, namely

$$\begin{bmatrix} dF_r(t,z) \\ dF_t(t,z) \\ dF_a(t,z) \end{bmatrix} = \sum_{j=1}^{N} g(\varphi_j(t,z)) \begin{bmatrix} dF_{r,j}(t,z) \\ dF_{t,j}(t,z) \\ dF_{a,j}(t,z) \end{bmatrix} \tag{5.2}$$

The cutting force includes the shear force acting on the rake surface and the edge force produced by the ploughing action between the flank face and the machined surface, so the cutting force can be decomposed into the following forms:

$$\begin{bmatrix} dF_r(t,z) \\ dF_t(t,z) \\ dF_a(t,z) \end{bmatrix} = \sum_{j=1}^{N} g(\varphi_j(t,z)) \begin{bmatrix} dF^c_{r,j}(t,z) + dF^e_{r,j}(t,z) \\ dF^c_{t,j}(t,z) + dF^e_{t,j}(t,z) \\ dF^c_{a,j}(t,z) + dF^e_{a,j}(t,z) \end{bmatrix} \tag{5.3}$$

The cutting force acted on the tool during milling is closely related to the chip thickness. In the above cutting force model, the cutting force can be expressed as a function of chip thickness and cutting force coefficient. According to the linear

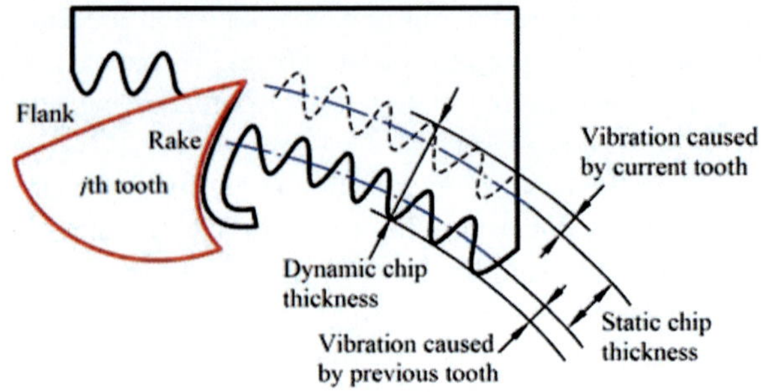

Fig. 5.2 Static and dynamic chip thickness in milling

edge force theory, the instantaneous static and dynamic cutting forces of tooth j at the axial height z are expressed as:

$$\begin{bmatrix} dF_{r,j}^{cs}(t,z) \\ dF_{t,j}^{cs}(t,z) \\ dF_{a,j}^{cs}(t,z) \end{bmatrix} = \begin{bmatrix} K_{rc} \\ K_{tc} \\ K_{ac} \end{bmatrix} h_{s,j}(t,z)ds \tag{5.4}$$

$$\begin{bmatrix} dF_{r,j}^{cd}(t,z) \\ dF_{t,j}^{cd}(t,z) \\ dF_{a,j}^{cd}(t,z) \end{bmatrix} = \begin{bmatrix} K_{rc} \\ K_{tc} \\ K_{ac} \end{bmatrix} h_{d,j}(t,z)ds \tag{5.5}$$

where $h_{s,j}(t,z)$ is the static chip thickness, $h_{d,j}(t,z)$ is the dynamic chip thickness.

In the low rigidity machining system, due to the insufficient rigidity of the tool and workpiece, the relative displacement between them will generate under the action of the cutting force, and this displacement generally occurs in the main vibration direction. The existence of vibration and relative displacement causes the fluctuation of the machined surface texture, which results in the variation of the dynamic chip thickness during the process, as shown in Fig. 5.2. Therefore, the instantaneous chip thickness in the milling process includes two parts: the static chip thickness corresponding to the feed rate and the dynamic chip thickness caused by the relative vibration between the tool and the workpiece.

In the milling with small cutting depth, due to the intermittent cutting of the tool and the fluctuation of chip thickness, there will be two alternating cutting forces with different mechanisms. One is the alternating component of cutting force caused by the regeneration effect of the tool which could cause unsteady chatter during the machining process. The other is the alternating component of cutting force caused by the intermittent cutting of the tool teeth in the state of no chatter. This periodic impact will lead to periodic forced vibration. Therefore, the alternating excitation force in milling can be further decomposed into dynamic cutting force and static cutting force, as shown in Fig. 5.3.

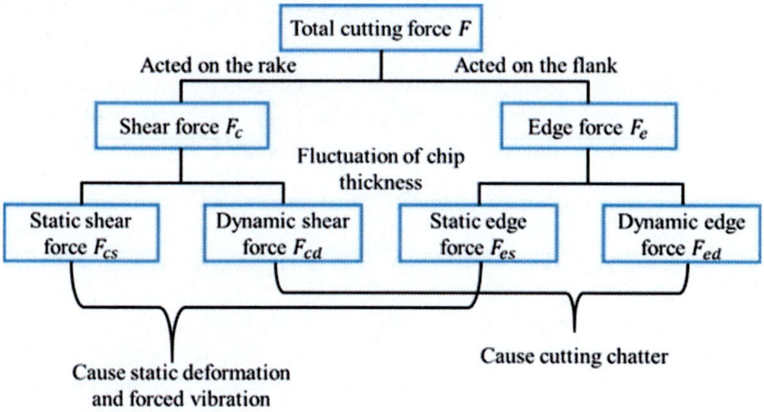

Fig. 5.3 Alternating excitation force decomposition diagram

5.3 Prediction of Milling Dynamic Response

5.3.1 Forced Vibration in Milling

In the milling process of thin-walled parts, with the continuous removal of the workpiece material, the cutting force and the modal parameters of the machining system will also change. Even when the machining system is in a stable state, some parts of the workpiece will still exist unpredictable vibration. The reason of this phenomenon is that the rotation of the tool causes the teeth to continuously cut in and out of the workpiece, and this intermittent cutting brings continuous impact to the workpiece, which induces forced vibration. Forced vibration has a significant impact on machining errors. Generally speaking, forced vibration in the milling process almost always exists although there is no chatter, in most cases, it is not very serious compared to the chatter during unsteady milling. However, for parts with small wall thickness, the forced vibration shows similar surface quality damage compared with chatter.

The intermittent cutting of the tool teeth will produce periodic impacts, which always exist in the process system as a forced excitation, thereby forming a forced vibration system. In the case of no chatter, the dynamic cutting force in the process system is zero, and only the periodic static cutting force needs to be considered. For the thin-walled parts milling systems, the forced excitation force can be expressed as a harmonic force:

$$G(t) = G_0\cos(\omega t) \tag{5.6}$$

Referring to Eq. (2.1), the forced vibration differential equation of the machining process system instability mode is expressed as:

$$m\ddot{x}(t) + c\dot{x}(t) + kx(t) = G_0\cos(\omega t) \tag{5.7}$$

Therefore, the system will experience forced vibration with the same frequency as the excitation force, but having time or phase lag. Assuming that the transient vibration caused by the initial load has disappeared and the system is in a stable state, then

$$x(t) = X\sin(\omega t + \varphi) \tag{5.8}$$

where X is the amplitude of forced vibration, φ is the phase angle.

Using the complex harmonic function to express forced vibration will make the corresponding mathematical calculations more convenient, then the harmonic force and response can be described as:

$$G(t) = G_0 e^{j\alpha} e^{j\omega t} \tag{5.9}$$

$$x(t) = Xe^{j(\omega t + \varphi)} \tag{5.10}$$

where α is the phase advance relative to the reference time or angle on the complex plane.

Substitute Eqs. (5.9) and (5.10) into forced vibration differential Eq. (5.7), it can be derived as:

$$\left(k - \omega^2 m + j\omega c\right) X e^{j\varphi} e^{j\omega t} = G_0 e^{j\alpha} e^{j\omega t} \tag{5.11}$$

Then the amplitude and phase angle of resonance are:

$$|\Phi(\omega)| = \left|\frac{X}{G_0}\right| = \frac{1}{k\sqrt{(1-\gamma^2)^2 + (2\xi\gamma)^2}} \tag{5.12}$$

$$\varphi = \arctan\left(\frac{-2\xi\gamma}{1-\gamma^2}\right) + \alpha \tag{5.13}$$

where γ is the natural frequency ratio, ξ is the damping ratio. Equations (5.12) and (5.13) is the transfer function or frequency response function of the process system.

Define the magnification ratio of the amplitude of forced vibration relative to the static deformation X_0, denoted by β, which reflects the severity of the forced vibration.

$$\beta = \frac{X}{X_0} = \frac{1}{\sqrt{(1-\gamma^2)^2 + (2\xi\gamma)^2}} \tag{5.14}$$

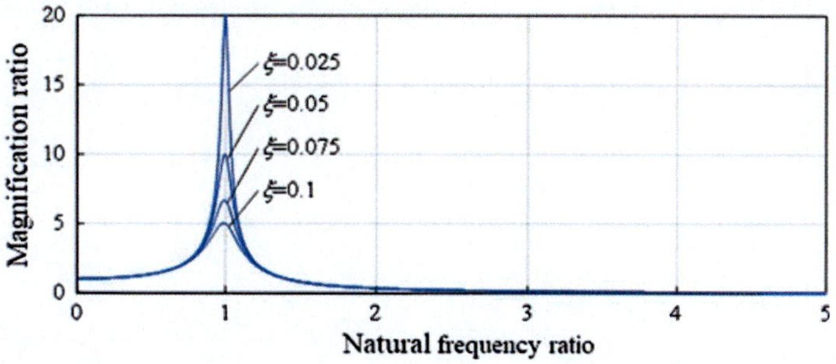

Fig. 5.4 The relationship between the natural frequency ratio and the magnification ratio

For different damping ratio ξ, according to the forced vibration theory, the relationship between the natural frequency ratio and the magnification ratio can be obtained, as shown in Fig. 5.4.

According to the definition of tooth passing frequency, the relationship between spindle speed and magnification ratio can be expressed as:

$$\beta = \frac{X}{X_0} = \frac{1}{\sqrt{\left[1 - \left(\frac{nN}{60f_d}\right)^2\right]^2 + \left(\frac{\xi nN}{30f_d}\right)^2}} \tag{5.15}$$

In an ideal situation, when the tool and workpiece are rigid, the relative movement of the tool relative to the workpiece in the feed direction is recorded as $u_r(t)$, which is related to the feed rate. In the case of no chatter, when considering the flexibility of the workpiece, the actual movement between the two is composed of the ideal movement and the movement caused by the system vibration, which can be expressed as:

$$u(t) = u_r(t) + u_s(t) \tag{5.16}$$

Then under the action of periodic cutting force, the error ε_{SLE} between the designed surface and the actual machined surface caused by the vibration of the machining process system is:

$$\varepsilon_{\text{SLE}} = \text{sign}[u_r(t) - u(t)] \tag{5.17}$$

where $\text{sign}(\cdot)$ is used to distinguish down milling and up milling, taking -1 when down milling and 1 when up milling. The machining error varying with the spindle speed obtained by simulation is shown in Fig. 5.5.

From Figs. 5.4 and 5.5, it can be found that when the tooth passing frequency is close to the natural frequency of the machining process system (i.e., the natural

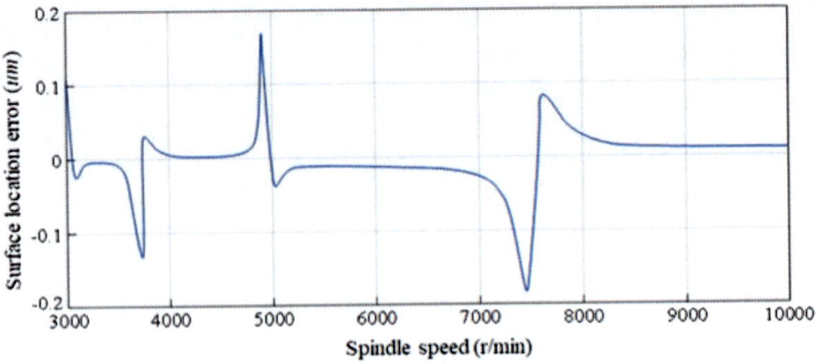

Fig. 5.5 The relationship between the surface location error and spindle speed

frequency ratio γ approaches 1), the amplitude of forced vibration increases significantly under the action of periodic excitation. It can be seen from the definition of the tooth passing frequency that by adjusting the spindle speed, we can control γ and then change β, thereby limiting the amplitude of the forced vibration caused by the alternating exciting force at a certain range to avoid excessive vibration which affects the surface quality of thin-walled parts and aggravates tool wear.

5.3.2 *Prediction of Milling Chatter Stability*

5.3.2.1 Efficient Fully Discrete Prediction Method for Chatter Stability

The model equation for the dynamic milling process with a single time delay can be expressed in the state-space form:

$$\dot{\boldsymbol{x}}(t) = \boldsymbol{A}\boldsymbol{x}(t) + \boldsymbol{B}(t)\boldsymbol{x}(t) - \boldsymbol{B}(t)\boldsymbol{x}(t-\tau) \tag{5.18}$$

where $\boldsymbol{A}$ is a constant matrix, $\boldsymbol{B}(t)$ is a time-periodic matrix that satisfies $\boldsymbol{B}(t) = \boldsymbol{B}(t+\tau)$, τ is the time delay, and $\boldsymbol{x}(t)$ represents the relative displacement between the tool and the workpiece.

To calculate state equations, i.e., Eq. (5.18), the terms $\boldsymbol{B}(t)\boldsymbol{x}(t)$ and $\boldsymbol{B}(t)\boldsymbol{x}(t-\tau)$ on the right side are approximated, while the term $\dot{\boldsymbol{x}}(t)$ on the left side stays unchanged. Then the delayed differential equations can be expressed as a series of ordinary differential equations. Firstly, the time period was divided into m time-intervals (with each interval length $\Delta t = \tau/m$), for jth time step $[t_j, t_{j+1}]$, the ordinary differential equation can be represented as:

$$\dot{\boldsymbol{x}}(t) = \boldsymbol{A}\boldsymbol{x}(t) + \tilde{\boldsymbol{B}}(t)\tilde{\boldsymbol{x}}(t) - \tilde{\boldsymbol{B}}(t)\tilde{\boldsymbol{x}}(t-\tau), t \in [t_j, t_{j+1}] \tag{5.19}$$

where $t_j = j\Delta t$, $j = 1, 2, \ldots, m$. The derivatives of $\boldsymbol{x}(t)$ at $t = t_j$ and $t = t_{j+1}$ could be obtained as:

$$\begin{cases} \dot{\boldsymbol{x}}(t_j) = \boldsymbol{A}\boldsymbol{x}(t_j) + \boldsymbol{B}(t_j)\left[\boldsymbol{x}(t_j) - \boldsymbol{x}(t_j - \tau)\right] \\ \dot{\boldsymbol{x}}(t_{j+1}) = \boldsymbol{A}\boldsymbol{x}(t_{j+1}) + \boldsymbol{B}(t_{j+1})\left[\boldsymbol{x}(t_{j+1}) - \boldsymbol{x}(t_{j+1} - \tau)\right] \end{cases} \tag{5.20}$$

From Eq. (5.20), it is obvious that in each time interval, $\dot{\boldsymbol{x}}(t_j)$ and $\dot{\boldsymbol{x}}(t_{j+1})$ could be expressed exactly with matrixes $\boldsymbol{A}$, $\boldsymbol{B}(t_j)$, $\boldsymbol{B}(t_{j+1})$ and several discrete state items. By utilizing the two derivatives and the state items $\boldsymbol{x}(t_j)$ and $\boldsymbol{x}(t_{j+1})$, then the value $\boldsymbol{x}(t)$ in the time span $[t_j, t_{j+1}]$ could be approximated by using Hermite interpolation as:

$$\tilde{\boldsymbol{x}}(t) = \boldsymbol{K}_1(t)\boldsymbol{x}_j + \boldsymbol{K}_2(t)\boldsymbol{x}_{j+1} + \boldsymbol{K}_3(t)\boldsymbol{x}_{j-r} + \boldsymbol{K}_4(t)\boldsymbol{x}_{j-r+1} \tag{5.21}$$

where $\boldsymbol{x}_j$ represents $\boldsymbol{x}(j\Delta t)$ and

$$\begin{cases} \boldsymbol{K}_1(t) = \frac{(t-\Delta t)^2\left\{\left[\boldsymbol{I}+(\boldsymbol{A}+\boldsymbol{B}_j)t\right]\Delta t+2t\boldsymbol{I}\right\}}{\Delta t^3} \\ \boldsymbol{K}_2(t) = \frac{\left\{(-\boldsymbol{A}-\boldsymbol{B}_{j+1})\Delta t^2+\left[3\boldsymbol{I}+(\boldsymbol{A}+\boldsymbol{B}_{j+1})t\right]\Delta t-2t\boldsymbol{I}\right\}t^2}{\Delta t^3} \\ \boldsymbol{K}_3(t) = -\frac{(t-\Delta t)^2 t\boldsymbol{B}_j}{\Delta t^2} \\ \boldsymbol{K}_4(t) = -\frac{t^2(t-\Delta t)\boldsymbol{B}_{j+1}}{\Delta t^2} \end{cases} \tag{5.22}$$

As for the time-delayed item $\boldsymbol{x}(t - \tau)$ in the time span $[t_j, t_{j+1}]$, it could be represented by $\boldsymbol{x}_{j-m}$, $\boldsymbol{x}_{j-m+1}$ and $\boldsymbol{x}_{j-m+2}$, which is

$$\tilde{\boldsymbol{x}}(t - \tau) = \boldsymbol{K}_5(t)\boldsymbol{x}_{j-m} + \boldsymbol{K}_6(t)\boldsymbol{x}_{j-m+1} + \boldsymbol{K}_7(t)\boldsymbol{x}_{j-m+2} \tag{5.23}$$

where

$$\begin{cases} \boldsymbol{K}_5(t) = \left(\frac{1}{2}\frac{t^2}{\Delta t^2} - \frac{3}{2}\frac{t}{\Delta t} + 1\right)\boldsymbol{I} \\ \boldsymbol{K}_6(t) = \left(-\frac{t^2}{\Delta t^2} + \frac{2t}{\Delta t}\right)\boldsymbol{I} \\ \boldsymbol{K}_7(t) = \left(\frac{1}{2}\frac{t^2}{\Delta t^2} - \frac{1}{2}\frac{t}{\Delta t}\right)\boldsymbol{I} \end{cases} \tag{5.24}$$

Similarly, during the time period $[t_j, t_{j+1}]$, $\tilde{\boldsymbol{B}}(t)$ can be easily expressed by $\boldsymbol{B}_j$ and $\boldsymbol{B}_{j+1}$ as:

$$\tilde{\boldsymbol{B}}(t) = \boldsymbol{B}_j + \frac{\boldsymbol{B}_{j+1} - \boldsymbol{B}_j}{\Delta t}(t - t_j) \tag{5.25}$$

Then in the time period $[t_j, t_{j+1}]$, Eq. (5.19) can be solved as:

$$\boldsymbol{Q}_j\boldsymbol{y}_{j+1} = \boldsymbol{H}_j\boldsymbol{y}_j + \boldsymbol{H}_{j-m}\boldsymbol{y}_{j-m} + \boldsymbol{H}_{j-m+1}\boldsymbol{y}_{j-m+1} + \boldsymbol{H}_{j-m+2}\boldsymbol{y}_{j-m+2} \tag{5.26}$$

where $\boldsymbol{Q}_j$, $\boldsymbol{H}_j$, $\boldsymbol{H}_{j-m}$, $\boldsymbol{H}_{j-m+1}$ and $\boldsymbol{H}_{j-m+2}$ are the coefficient matrixes that are relevant to the time-periodic matrix $\boldsymbol{B}_j$ and $\boldsymbol{B}_{j+1}$. And the expressions are:

$$\begin{aligned}
\boldsymbol{Q}_j &= \boldsymbol{I} - \boldsymbol{d}_1\boldsymbol{B}_j - \boldsymbol{d}_2\boldsymbol{B}_{j+1}\\
\boldsymbol{H}_j &= \boldsymbol{s}_1\boldsymbol{B}_j + \boldsymbol{s}_2\boldsymbol{B}_{j+1} + e^{\boldsymbol{A}\Delta t}\\
\boldsymbol{H}_{j-m} &= \boldsymbol{e}_1\boldsymbol{B}_j + \boldsymbol{e}_2\boldsymbol{B}_{j+1}\\
\boldsymbol{H}_{j-m+1} &= \boldsymbol{f}_1\boldsymbol{B}_j + \boldsymbol{f}_2\boldsymbol{B}_{j+1}\\
\boldsymbol{H}_{j-m+2} &= \boldsymbol{g}_1\boldsymbol{B}_j + \boldsymbol{g}_2\boldsymbol{B}_{j+1}
\end{aligned}$$

where $\boldsymbol{I}$ is the identity matrix, $\boldsymbol{d}_1$ and $\boldsymbol{d}_2$ are constant matrixes. The coefficient matrixes $\boldsymbol{d}_1, \boldsymbol{d}_2, \boldsymbol{s}_1, \boldsymbol{s}_2, \boldsymbol{e}_1, \boldsymbol{e}_2, \boldsymbol{f}_1, \boldsymbol{f}_2, \boldsymbol{g}_1, \boldsymbol{g}_2$ are represented as:

$$\boldsymbol{d}_1 = -\left(\tfrac{\boldsymbol{A}}{\Delta t^3} - \tfrac{2\boldsymbol{I}}{\Delta t^4}\right)\boldsymbol{\Phi}_4 + \left(\tfrac{2\boldsymbol{A}}{\Delta t^2} - \tfrac{5\boldsymbol{I}}{\Delta t^3}\right)\boldsymbol{\Phi}_3 + \left(\tfrac{3\boldsymbol{I}}{\Delta t^2} - \tfrac{\boldsymbol{A}}{\Delta t}\right)\boldsymbol{\Phi}_2$$

$$\boldsymbol{d}_2 = \left(\tfrac{\boldsymbol{A}}{\Delta t^3} - \tfrac{2\boldsymbol{I}}{\Delta t^4}\right)\boldsymbol{\Phi}_4 + \left(\tfrac{3\boldsymbol{I}}{\Delta t^3} - \tfrac{\boldsymbol{A}}{\Delta t^2}\right)\boldsymbol{\Phi}_3$$

$$\boldsymbol{s}_1 = \left(\frac{5\boldsymbol{I}}{\Delta t^3} + \frac{3\boldsymbol{A}}{\Delta t^2}\right)\boldsymbol{\Phi}_3 + \boldsymbol{\Phi}_0 + \left(\boldsymbol{A} - \frac{\boldsymbol{I}}{\Delta t}\right)\boldsymbol{\Phi}_1 + \left(-\frac{3\boldsymbol{A}}{\Delta t} - \frac{3\boldsymbol{I}}{\Delta t^2}\right)\boldsymbol{\Phi}_2 - \left(\frac{2\boldsymbol{I}}{\Delta t^4} + \frac{\boldsymbol{A}}{\Delta t^3}\right)\boldsymbol{\Phi}_4$$

$$\boldsymbol{s}_2 = \frac{\boldsymbol{\Phi}_1}{\Delta t} + \frac{\boldsymbol{A}\boldsymbol{\Phi}_2}{\Delta t} + \left(-\frac{2\boldsymbol{A}}{\Delta t^2} - \frac{3\boldsymbol{I}}{\Delta t^3}\right)\boldsymbol{\Phi}_3 + \left(\frac{2\boldsymbol{I}}{\Delta t^4} + \frac{\boldsymbol{A}}{\Delta t^3}\right)\boldsymbol{\Phi}_4$$

$$\begin{aligned}
\boldsymbol{e}_1 &= \frac{1}{2}\frac{\boldsymbol{\Phi}_3}{\Delta t^3} - \boldsymbol{\Phi}_0 + \frac{5}{2}\frac{\boldsymbol{\Phi}_1}{\Delta t} - \frac{2\boldsymbol{\Phi}_2}{\Delta t^2}\\
\boldsymbol{e}_2 &= -\frac{\boldsymbol{\Phi}_1}{\Delta t} + \frac{3}{2}\frac{\boldsymbol{\Phi}_2}{\Delta t^2} - \frac{1}{2}\frac{\boldsymbol{\Phi}_3}{\Delta t^3}
\end{aligned}$$

$$\begin{aligned}
\boldsymbol{f}_1 &= -\frac{\boldsymbol{\Phi}_3}{\Delta t^3} + \frac{3\boldsymbol{\Phi}_2}{\Delta t^2} - \frac{2\boldsymbol{\Phi}_1}{\Delta t}\\
\boldsymbol{f}_2 &= \frac{\boldsymbol{\Phi}_3}{\Delta t^3} - \frac{2\boldsymbol{\Phi}_2}{\Delta t^2}
\end{aligned}$$

$$\begin{aligned}
\boldsymbol{g}_1 &= \frac{1}{2}\frac{\boldsymbol{\Phi}_3}{\Delta t^3} - \frac{\boldsymbol{\Phi}_2}{\Delta t^2} + \frac{1}{2}\frac{\boldsymbol{\Phi}_1}{\Delta t}\\
\boldsymbol{g}_2 &= -\frac{1}{2}\frac{\boldsymbol{\Phi}_3}{\Delta t^3} + \frac{1}{2}\frac{\boldsymbol{\Phi}_2}{\Delta t^2}
\end{aligned}$$

where

$$\boldsymbol{\Phi}_0 = \int_0^{\Delta t} e^{\boldsymbol{A}(\Delta t - \xi)} d\xi$$

$$\boldsymbol{\Phi}_1 = \boldsymbol{A}^{-1}(\boldsymbol{\Phi}_0 - \Delta t\boldsymbol{I})$$

$$\boldsymbol{\Phi}_2 = \boldsymbol{A}^{-1}\left(2\boldsymbol{\Phi}_1 - \Delta t^2 \boldsymbol{I}\right)$$
$$\boldsymbol{\Phi}_3 = \boldsymbol{A}^{-1}\left(3\boldsymbol{\Phi}_2 - \Delta t^3 \boldsymbol{I}\right)$$
$$\boldsymbol{\Phi}_4 = \boldsymbol{A}^{-1}\left(4\boldsymbol{\Phi}_3 - \Delta t^4 \boldsymbol{I}\right)$$

According to Eq. (5.26), a 2(m + 1) dimensional discrete map can be expressed as:

$$\boldsymbol{Z}_{j+1} = \boldsymbol{D}_j \boldsymbol{Z}_j \tag{5.27}$$

where $\boldsymbol{Z}_j = \begin{bmatrix} \boldsymbol{y}_j & \boldsymbol{y}_{j-1} & \boldsymbol{y}_{j-2} & \cdots & \boldsymbol{y}_{j-m} \end{bmatrix}^T$.

It is clear that if matrix $\boldsymbol{Q}$ is nonsingular, then $\boldsymbol{D}_j$ can be expressed as:

$$\boldsymbol{D}_j = \begin{bmatrix} \boldsymbol{Q}^{-1}\boldsymbol{H}_j & \boldsymbol{0} & \cdots & \boldsymbol{Q}^{-1}\boldsymbol{H}_{j-m+2} & \boldsymbol{Q}^{-1}\boldsymbol{H}_{j-m+1} & \boldsymbol{Q}^{-1}\boldsymbol{H}_{j-m} \\ \boldsymbol{I} & \boldsymbol{0} & \cdots & \boldsymbol{0} & \boldsymbol{0} & \boldsymbol{0} \\ \boldsymbol{0} & \boldsymbol{I} & \cdots & \boldsymbol{0} & \boldsymbol{0} & \boldsymbol{0} \\ \vdots & \vdots & \cdots & \vdots & \vdots & \vdots \\ \boldsymbol{0} & \boldsymbol{0} & \cdots & \boldsymbol{0} & \boldsymbol{I} & \boldsymbol{0} \end{bmatrix} \tag{5.28}$$

The approximate Floquet transition matrix is obtained as:

$$\boldsymbol{\Phi} = \boldsymbol{D}_{m-1}\boldsymbol{D}_{m-2}\cdots\boldsymbol{D}_0 \tag{5.29}$$

The stability of the system can be determined by judging whether the eigenvalues of the transition matrix $\boldsymbol{\Phi}$ are in modulus less than 1or not. If not, the system will be unstable, otherwise, it will be stable.

5.3.2.2 Rapid Prediction of Chatter Stability for Annular Casing End Milling

In the machining of complex aviation parts, the tools used are generally complicated, and the chatter stability prediction also has its own characteristics. For example, when processing annular casings and compressor blades of aeroengines, bull-nose end mills are commonly adopted, and the annular cutting edges are mainly engaged in cutting. Taking annular casing as an example, the semi-finishing and finishing process of the outer surface of the casing can be simplified into the bull-nose end machining with a fixed axis, as shown in Fig. 5.6.

In the milling process of the outer surface of the casing, the cutting forces excite the workpiece and then cause dynamic displacements which leads to dynamic chip thickness between the previous and present edge.

$$h_{d,j}(\varphi_j, \gamma) = (\Delta r \sin\gamma - \Delta z \cos\gamma) g(\varphi_j) \tag{5.30}$$

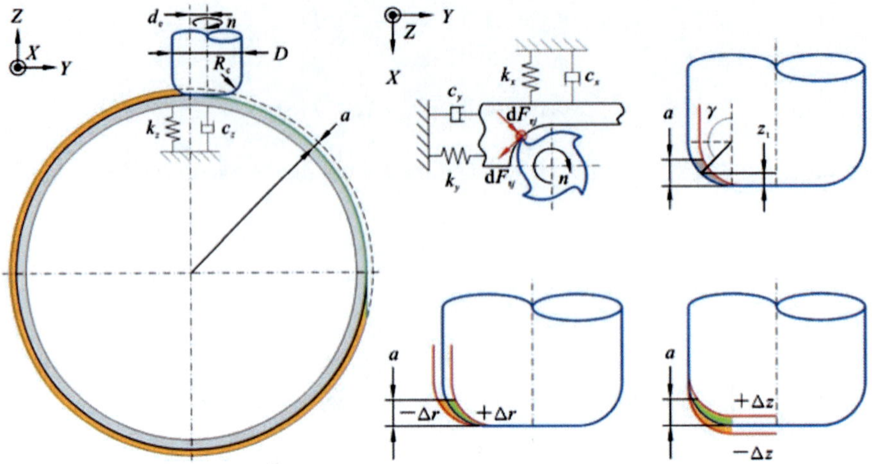

Fig. 5.6 Dynamic milling model for bull-nose end milling of aero-engine casings

with,

$$\Delta r = \Delta x \sin\varphi_j + \Delta y \cos\varphi_j \tag{5.31}$$

According to the 3D chatter stability model proposed by Altintas and Budak, the dynamic force acting on the cutter can be expressed as the following form in the time domain:

$$\begin{bmatrix} F_x(t) \\ F_y(t) \\ F_z(t) \end{bmatrix} = \frac{N}{4\pi} a K_t \boldsymbol{A}_1 \begin{bmatrix} \Delta x(t) \\ \Delta y(t) \\ \Delta z(t) \end{bmatrix} \tag{5.32}$$

where $F_x(t)$, $F_y(t)$ and $F_z(t)$ are the cutting forces along the x, y and z directions, respectively, $\Delta x(t)$, $\Delta y(t)$ and $\Delta z(t)$ are the dynamic displacements in the x, y and z directions, $\boldsymbol{A}_1$ is the matrix of directional coefficients.

Using the transfer function at the cutter-workpiece contact area, the dynamic displacements can be described in the frequency domain:

$$\begin{bmatrix} \Delta x(i\omega_c) \\ \Delta y(i\omega_c) \\ \Delta z(i\omega_c) \end{bmatrix} = (1 - e^{-i\omega_c T}) \boldsymbol{G} \begin{bmatrix} F_x(i\omega_c) \\ F_y(i\omega_c) \\ F_z(i\omega_c) \end{bmatrix} e^{i\omega_c T} \tag{5.33}$$

where ω_c is the chatter frequency, T is the tooth passing frequency, $\boldsymbol{G} = \boldsymbol{G}_c + \boldsymbol{G}_w$ is the sum of the transfer function matrixes of the cutter and the workpiece.

Therefore, the dynamic cutting force in the frequency domain can be expressed as:

$$\begin{bmatrix} F_x(i\omega_c) \\ F_y(i\omega_c) \\ F_z(i\omega_c) \end{bmatrix} = \frac{N}{4\pi} a K_t (1 - e^{-i\omega_c T}) \boldsymbol{A}_1 \boldsymbol{G} \begin{bmatrix} F_x(i\omega_c) \\ F_y(i\omega_c) \\ F_z(i\omega_c) \end{bmatrix} \tag{5.34}$$

Then the chatter stability prediction becomes an eigenvalue problem, where Eq. (5.35) defines the eigenequation and Eq. (5.36) defines the eigenvalue:

$$det\{\boldsymbol{I} + \lambda \boldsymbol{A}_1 \boldsymbol{G}\} = 0 \tag{5.35}$$

$$\lambda = -\frac{N}{4\pi} a K_t (1 - e^{-i\omega_c T}) \tag{5.36}$$

Therefore, the critical axial depth of cut and spindle speed in the milling process is:

$$a_{\lim} = -\frac{2\pi Re(\lambda)}{N K_t} \left[1 + \left(\frac{Im(\lambda)}{Re(\lambda)} \right)^2 \right] \tag{5.37}$$

$$n = \frac{60\omega_c}{N\left[(2k+1)\pi - 2\arctan\frac{Im(\lambda)}{Re(\lambda)}\right]} \tag{5.38}$$

where $Im(\lambda)$ is the imaginary part of λ, $Re(\lambda)$ is the real part of λ, k is the sequence number of lobes.

In the process of semi-finishing and finishing of the outer surface of the casing, the rotation of the workpiece and the tool realize the feed movement and the material cutting respectively. Generally, multi-axis processing is used to avoid cutting material with the bottom edge of the tool. During the bull-nose end milling, the smaller the lead angle is, the bigger the cutting width. In addition, a small lead angle can meet the needs of tool-axis directions because the outer surfaces of aeroengine casings have relatively open spaces. The annular, thin-walled structure of the casing makes the main mode of the workpiece subsystem along the surface normal direction, which means it is the main vibration direction. When the lead angle is equal to a small value, the dynamic displacement in the surface normal direction can be approximated by the dynamic displacement in z direction. And the dynamic displacements in x and y directions can be ignored because they are minor compared to the dominant mode. Hence, the chatter stability model presented here can be reduced to a 1D model in the normal direction. The stability problem can be simplified into:

$$F_z(i\omega_c) = \frac{N}{2\pi} a_z K_t \alpha_z Re\left[G_z(i\omega_c)\right] F_z(i\omega_c) \tag{5.39}$$

where

$$\alpha_z = \varphi\left[-K_r(\cos 2\gamma + 1) + K_a \sin 2\gamma\right]_{\varphi_{st}}^{\varphi_{ex}} \tag{5.40}$$

$$Re\left[G_z(i\omega_c)\right] = \frac{1 - d_z^2}{k_z\left[(1 - d_z^2)^2 + (2\xi_z d_z)^2\right]} \tag{5.41}$$

$$d_z = \frac{\omega_c}{\omega_n}$$

where a_z is the axial cutting depth of the tool, k_z and ξ_z is the stiffness and damping ratio in the z direction, ω_n is the natural frequency.

When the lead angle is equal to 0 degree, the cutting depth a_z is equal to machining allowance a. However, in the multi-axis milling process, there will be a small lead angle. Then, the relationship between the axial depth of cut a_z and machining allowance a is expressed as:

$$a_z = \frac{a}{\cos\theta} \tag{5.42}$$

Equation (5.39) can be changed as:

$$F_z(i\omega_c) = \frac{N}{2\pi\cos\theta} a K_t \alpha_z Re\left[G_z(i\omega_c)\right] F_z(i\omega_c) \tag{5.43}$$

Because of the complex cutter and workpiece geometry, the engagement region is relatively complicated. Also, the cutting force coefficients and the start and exit angles are not constant along the tool-axis direction. Then a disk-based method is extended to calculate the stability limit in multi-axis milling of aero-engine casings. As shown in Fig. 5.7, each disk element has a height of Δa. The bull-nose end mill is divided into several disk elements along the tool-axis direction. In order to obtain the total stability limit of the process system, the dynamic cutting forces of all the disk elements have to be summed up:

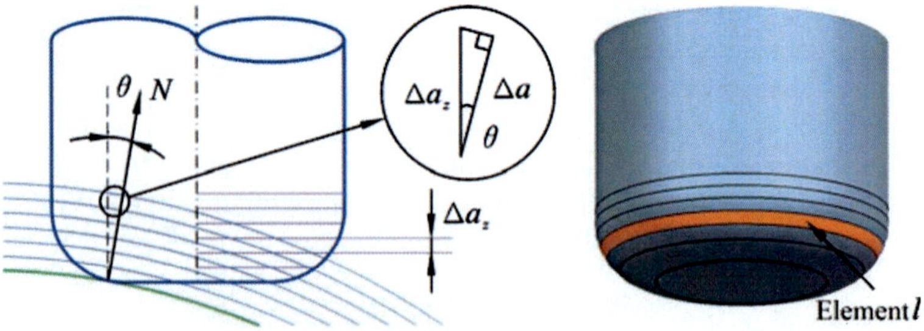

Fig. 5.7 Mechanics and dynamics of disk l

$$F_z(i\omega_c) = \frac{N}{2\pi\cos\theta}\Delta a\left(\sum_{l=1}^{m} K_{t,l}\alpha_{z,l}\right)Re\left[G_z(i\omega_c)\right]F_z(i\omega_c) \tag{5.44}$$

Then the stability limit and spindle speed corresponding to Δa are obtained:

$$\Delta a_{\lim} = \frac{2\pi\cos\theta}{N\sum_{l=1}^{m} K_{t,l}\alpha_{z,l}Re\left[G_z(i\omega_c)\right]} \tag{5.45}$$

$$n = \frac{30\omega_c}{N\left[(k+1)\pi - \arctan\frac{d_z^2-1}{2\xi_z d_z}\right]} \tag{5.46}$$

Because there are m disks in the calculation, the total stability limit for a given chatter frequency ω_c is calculated in the following form:

$$a_{\lim} = m\Delta a_{\lim} \tag{5.47}$$

5.4 Dynamic Response Control of Milling Based on Optimization of Cutting Parameters

In the milling process of thin-walled parts, the stability of the process system and surface quality are closely related to the system stiffness and cutting parameters such as spindle speed and axial cutting depth. At present, optimization of cutting parameters and increasing process rigidity are mainly used to control the dynamic response of the milling process. Optimization of cutting parameters is to select appropriate axial cutting depth-spindle speed parameter combination to avoid chatter and forced vibration based on the chatter Stability Lobe Diagram (SLD) and the surface position error diagram.

It can be seen from Chap. 2 that material removal in actual machining will cause significant changes to the dynamic characteristics of the process system, and lead to the staggered arrangement of the SLDs which makes it difficult to select cutting parameters. In order to solve these problems, optimization of cutting parameters for a single toolpath can be adopted, which can also be named as the fixed spindle speed path by path strategy. This method aims to establish a 3D SLD describing the dynamic evolution of the process system, and take material removal sequence in the single toolpath as the third dimension. Figures 5.8 and 5.9 show the 3D SLD and its 2D projection corresponding to the whole process and cutting row l_1 in casing machining. It can be found that in a single toolpath, the SLDs corresponding to the initial and final cutting segment together constitute the boundary of the entire cutting row. Because the dynamic characteristics of the process system within a single path have small changes, a larger stable parameter domain, a wider spindle speed range and a larger stability limit cutting depth can be determined by the 3D SLD. However,

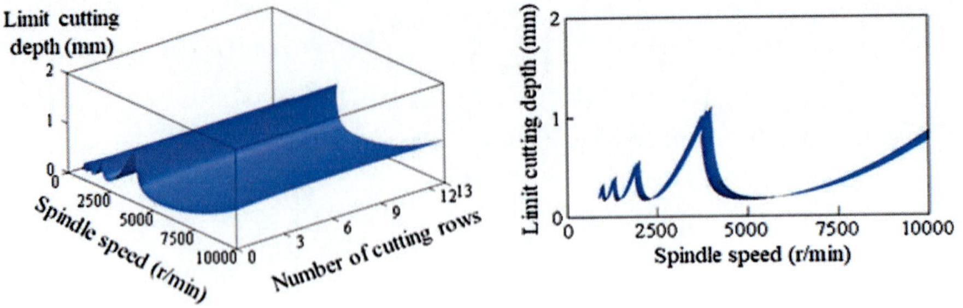

Fig. 5.8 3D stability lobe diagram and its 2D projection corresponding to the cutting row l_1

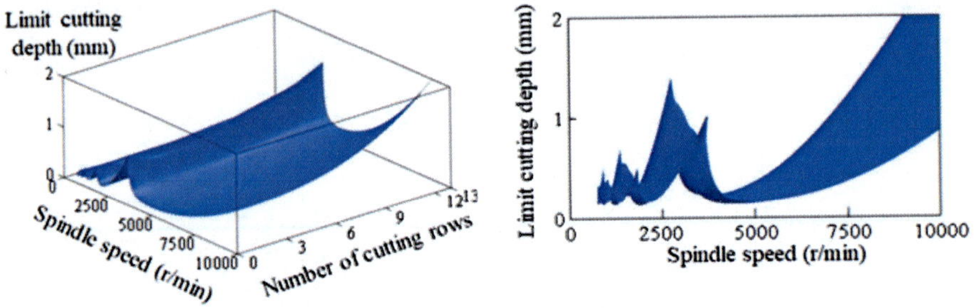

Fig. 5.9 3D stability lobe diagram and its 2D projection corresponding to the whole process

in the entire process, it will have significant limitations to acquire optimal cutting parameters when using 3D SLD and it is easy to result in low fault tolerance of dynamic response and poor machining efficiency.

It is possible to intuitively analyze the stability of the milling process in a single toolpath or the whole process if intercepting the section of the 3D SLD at different spindle speeds. Figure 5.10 shows the 3D SLD section at different spindle speeds corresponding to the cutting row l_1. It can be found that under the machining

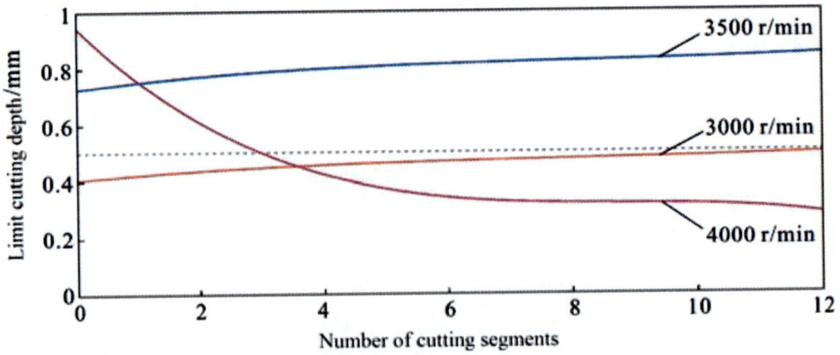

Fig. 5.10 3D stability lobe section at different spindle speeds corresponding to the cutting row l_1

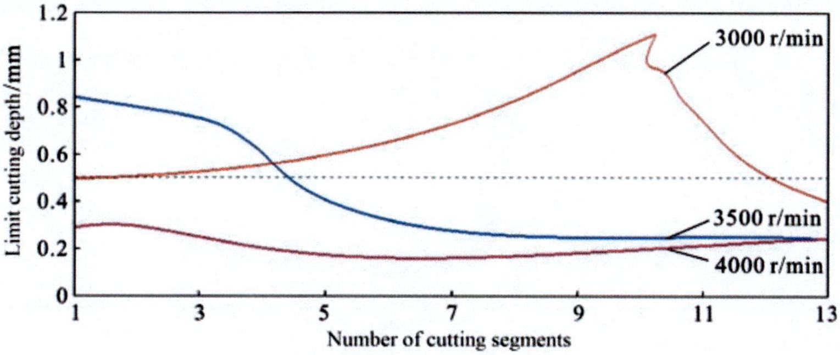

Fig. 5.11 Section of 3D stability lobe at different spindle speeds corresponding to the whole process

allowance of 0.5 mm, when the spindle speed is 3000 r/min, each cutting segment will be in chatter. When the spindle speed is 3500 r/min, each cutting segment is in a stable state. When the spindle speed is 4000 r/min, the previous cutting segments are in a stable state whereas subsequent cutting segments are in chatter. Figure 5.11 shows the 3D SLD section at different spindle speeds corresponding to the whole process. It can be found that when the spindle speed is 4000 r/min, each cutting row will be in chatter. When the spindle speed is 3000 r/min and 3500 r/min, some rows are in a chatter state, and others are in a stable state. Therefore, it is necessary to select different spindle speeds for different cutting rows to achieve stable cutting, thereby ensuring a stable state in the whole process. The change of the spindle speed can be realized in the entering and exiting of each cutting toolpath.

In addition, during the processing of the casing surface, if generating excessive alternating excitation force, the amplitude of forced vibration will exceed the allowable limit of the tool or workpiece, resulting in damage and scrapping to the tool or workpiece. Therefore, it is necessary to control the amplitude of forced vibration in the processing of the casing. When the tooth passing frequency is close to the natural frequency of the tool, the vibration amplitude of the process system will be enlarged, which will intensify the vibration of the machined surface, also increase the error. Therefore, surface location error can be introduced into the 3D SLD to optimize cutting parameters and achieve the purpose of suppressing chatter and forced vibration. As shown in Fig. 5.12, it can be found that there are four forced vibration amplitude amplification zones within the spindle speed range of 0–10000r/min, of which Zone 2, Zone 3, Zone 4 and Zone 5 are located near the spindle speed of 7404 r/min, 4936 r/min, 3702 r/min and 2962 r/min, respectively. The forced vibration amplitude amplification zones and the boundary of the SLD together constitute the constraint conditions of stable parameter area.

When selecting appropriate cutting parameters, on the one hand, it is necessary to avoid chatter based on the predicted stability lobe diagram, on the other hand, to avoid the forced vibration amplitude amplification area to control the surface position error.

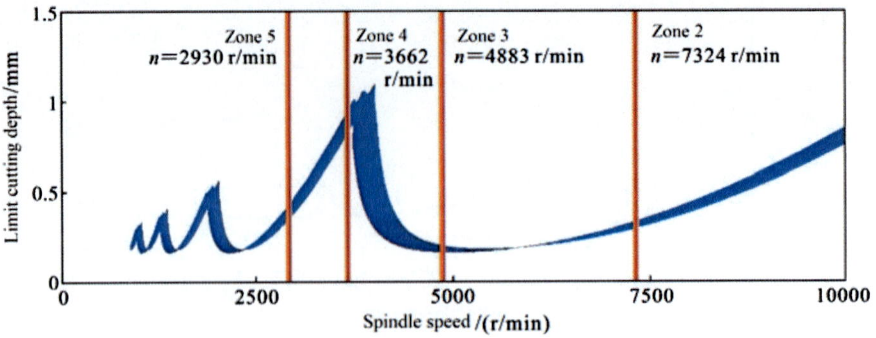

Fig. 5.12 Parameter optimization area of dynamic response control corresponding to the cutting row l_1

The constraints of spindle speed n are:

$$\begin{cases} 0 < n \le n_{\max} \\ n \ne \frac{60 f_{tpf}}{Nk} \end{cases} (k = 1, 2, \ldots) \tag{5.48}$$

where $n_{\max}$ is the allowable maximum spindle speed of the machine tool.

The constraints of axial cutting depth a_p are:

$$\begin{cases} 0 < a_p \le a_{p\max} \\ 0 < a_p \le a_{\lim}(n) \end{cases} \tag{5.49}$$

where $a_{p\max}$ is the maximum cutting depth of the process, $a_{\lim}(n)$ is the limit cutting depth when staying stable cutting, and it needs to be matched with spindle speed.

According to the above constraints, select optimal cutting parameters in the figure and simultaneously control chatter and forced vibration in a single toolpath. Figure 5.13 shows optimization results of cutting parameters corresponding to the cutting row l_1. The maximum allowable spindle speed of the machine tool is 8000 r/min and the selected cutting depth for semi-finishing is 0.5 mm. It can be found from Fig. 5.13 that there is no chatter in the spindle speed of 3715–3779 r/min, and forced vibration will occur when the selected spindle speed is near 3702 r/min. Select adjustment interval of spindle speed as 200 r/min, and select the middle value (3477 r/min) of the selectable range as the spindle speed used in the cutting row l_1. The spindle speed range 3377–3577 r/min does not cover Zone 4, and can simultaneously avoid chatter and forced vibration in the whole toolpath. The fault-tolerant cutting depth corresponding to the spindle speed of 3377–3577 r/min is 0.638 mm, that is, the allowable fluctuation range of the axial cutting depth is 0.138 mm. When the forced vibration amplitude amplification zone is located in the middle of the spindle speed allowable interval which causes difficult selection, it is better to select on the right side of the allowable interval to obtain a larger fault-tolerant cutting depth. At

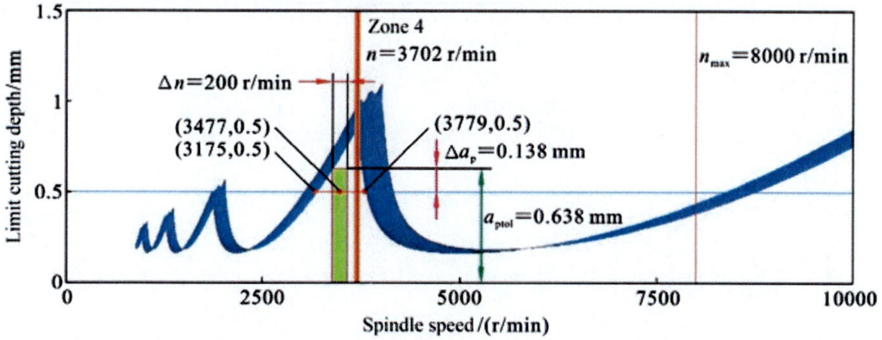

Fig. 5.13 Cutting parameters optimization result corresponding to the cutting row l_1

the same time, the adjustment interval of spindle speed can be appropriately reduced to ensure stable cutting.

5.5 Response Control Method Based on Variable Pitch Cutters Optimization Design

Conventional milling cutters generally adopt uniform pitch angle design. In recent years, some researchers have found that the variable pitch angle has a significant effect in suppressing the milling vibration. This section focuses on the design method of variable pitch cutters [3].

5.5.1 Stability Limit Calculation of Variable Pitch Cutters

As shown in Fig. 5.14, considering two degrees of freedom regeneration chatter model, due to the existence of relative cutter-workpiece displacements, the total chip thickness $h_j(t)$ consists of the static $S_j \sin\phi_j(t)$ (see Fig. 5.14d) and dynamic $\nu_j(t) - \nu_j(t-T)$ parts (see Fig. 5.14e and f). The resultant chip thickness $h_j(t)$ can be expressed as:

$$h_j(t) = S_j \sin\phi_j(t) + \nu_j(t) - \nu_j(t-T) \tag{5.50}$$

where $\nu_j(t)$ and $\nu_j(t-T)$ are the relative dynamic displacements of the current and previous tooth between the cutter and workpiece, respectively. The static component of the chip thickness can be neglected in the stability analysis since it does not affect the dynamic chip regeneration mechanism. Therefore, the dynamic chip thickness, which contributes to the dynamic chip load regeneration mechanism, can be described as:

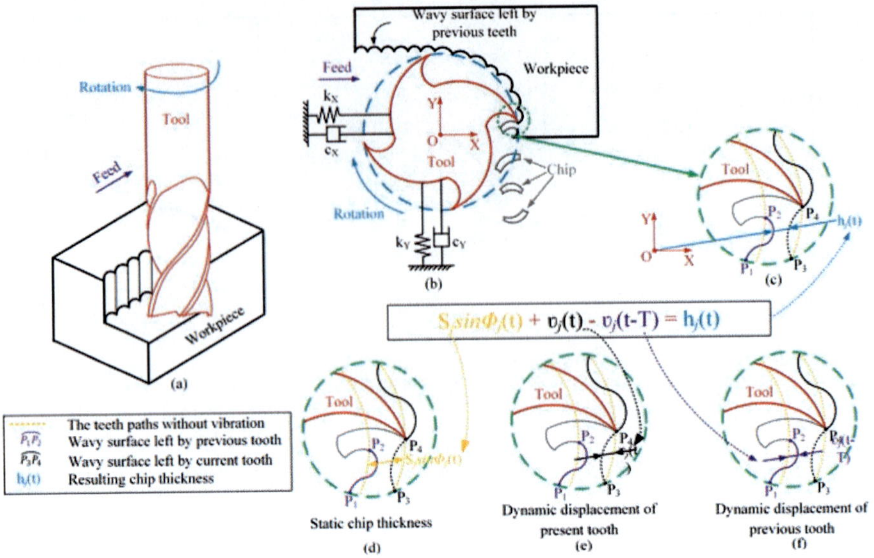

Fig. 5.14 Two degrees of freedom regeneration chatter model

$$h_j(t) = \Delta x \sin\phi_j(t) + \Delta y \cos\phi_j(t) \tag{5.51}$$

where Δx and Δy represent the difference of dynamic displacements between the present and previous tooth, respectively.

Thus, the dynamic milling forces can be derived as:

$$\begin{cases} F_{tj} = K_t a h_j(t) \\ F_{rj} = K_r F_{tj} \end{cases} \tag{5.52}$$

According to the Zero-Order Solution of chatter stability prediction [2], the limit stability cutting depth can be described as Eq. (5.37). For variable pitch cutters, the phase difference ε_j between the inner and outer waves corresponding to the pitch angle P_j is defined as:

$$\varepsilon_j = \omega_c T_j \tag{5.53}$$

where T_j and ε_j are the jth tooth period and phase difference, respectively. Therefore, the critical axial depth of cut for variable pitch cutter is:

$$a_{\lim}^{vp} = -\frac{4\pi Im(\lambda)}{K_t S} \tag{5.54}$$

where

$$S = \sum_{j=1}^{N} \sin(\omega_c T_j) = \sum_{j=1}^{N} \sin\varepsilon_j \tag{5.55}$$

Therefore, it could be observed from Eq. (5.54) that the variable pitch angles should be designed to minimize S for the purpose of increasing the stability limit (i.e., the critical axial depth of cut).

5.5.2 Geometrical Relation Between Adjacent Pitch Angles

For the cutters with alternating variable pitches, the pitch angles can be generally described as $P_1, P_2, P_1, P_2, \ldots$, where P_1 and P_2 represent the two different pitch angles of the cutter. Hence, for a cutter with N teeth, the alternating variable pitches should satisfy the following relation:

$$\sum_{i=1}^{N} P_i = 2\pi \Rightarrow \begin{cases} \frac{N(P_1+P_2)}{2} = 2\pi, & N \text{ is even} \\ [(N+1)P_1 + (N-1)P_2]/2 = 2\pi, & N \text{ is odd} \end{cases} \tag{5.56}$$

Considering that the tooth number of the variable pitch cutter N is an even value, the relationship between the two adjacent pitch angles can be expressed as:

$$P_1 + P_2 = \frac{4\pi}{N} \tag{5.57}$$

Based on Eq. (5.53), the phase differences corresponding to the two pitch angles are:

$$\begin{cases} \varepsilon_1 = \frac{60\omega_c P_1}{2\pi n} \\ \varepsilon_2 = \frac{60\omega_c P_2}{2\pi n} \end{cases} \tag{5.58}$$

Substituting Eq. (5.57) into Eq. (5.58), the following equation could be obtained:

$$\varepsilon_2 = \varepsilon_s - \varepsilon_1 \tag{5.59}$$

where $\varepsilon_s = 120\omega_c/(Nn)$.

Substituting Eq. (5.59) into Eq. (5.55), it can be derived as:

$$S = \frac{N[\sin\varepsilon_1 + \sin(\varepsilon_s - \varepsilon_1)]}{2} \tag{5.60}$$

For the cutter with fixed numbers of teeth, N is a constant value. Therefore, only the $\sin\varepsilon_1 + \sin(\varepsilon_s - \varepsilon_1)$ influences the value of S.

5.5.3 Design of Variable Pitch Angles

A new variable $\overline{S}$ is defined to as: $\overline{S} = \sin\varepsilon_1 + \sin(\varepsilon_s - \varepsilon_1)$. For the design of variable pitch milling cutters, minimizing the value of $\overline{S}$ is always the aim. The value of $\overline{S}$ is determined only by the values of ε_1 and ε_s, in which the value of ε_s depends only on given milling conditions. Variable pitch angle can only influence the value of ε_1. The challenge is that ε_s varies due to different machining systems and cutting conditions. The variation of ε_s should, therefore, be considered to ensure the robustness of the variable pitch design method whilst selecting a right/general value of ε_1 in order to increase the stability limit.

The variation of ε_1 and ε_s on $\overline{S}$ is shown in Fig. 5.15a. Because the value of $a_{\lim}^{vp}$ is positive, only considering the absolute value of $|\overline{S}|$ as shown in Fig. 5.15b. It can be seen that $\varepsilon_1 = 0, \pi, 2\pi$ result in the same minimum of $|\overline{S}|$. However, when $\varepsilon_1 = \pi$, the dynamic chip thickness and cutting forces will increase. Thus, $\varepsilon_1 = 0$ and $\varepsilon_1 = 2\pi$ are the general solutions. To be general, $\varepsilon_1 = 2k\pi$ (where k is a natural number) can be the solution with satisfactory robustness to increase the stability limit and reduce the cutting forces simultaneously.

Substituting $\varepsilon_1 = 2k\pi$ into Eq. (5.58), the optimized pitch angle is obtained as follows:

$$\begin{cases} P_1 = 2k\pi \times \frac{2\pi n}{60\omega_c}, k = 1, 2, 3, \ldots \\ P_2 = \frac{4\pi}{N} - P_1 \end{cases} \tag{5.61}$$

The combination of a set of four-fluted variable pitch angles for a certain process system designed by the above method is (84°, 96°, 84°, 96°). Compared with the uniform pitch angle of 90°, the corresponding chatter stability analysis is shown in Fig. 5.16. It can be seen that: (1) when the cutting depth was 1 mm, for the uniform pitch cutter, the spindle speed only could be selected from the range of r_{u1}, r_{u2}, r_{u3} and r_{u4}. For the variable pitch cutter, the stable spindle speed area was expanded to

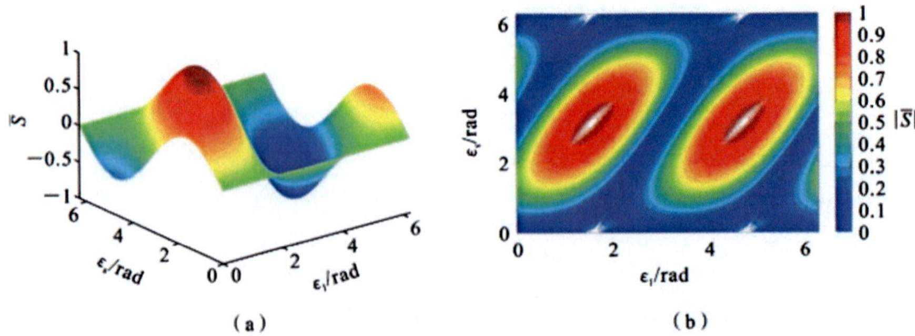

Fig. 5.15 The variation of ε_1 and ε_s on $\overline{S}$. **a** Three-dimensional surface of $\overline{S}$. **b** Two-dimensional contour of $\overline{S}$

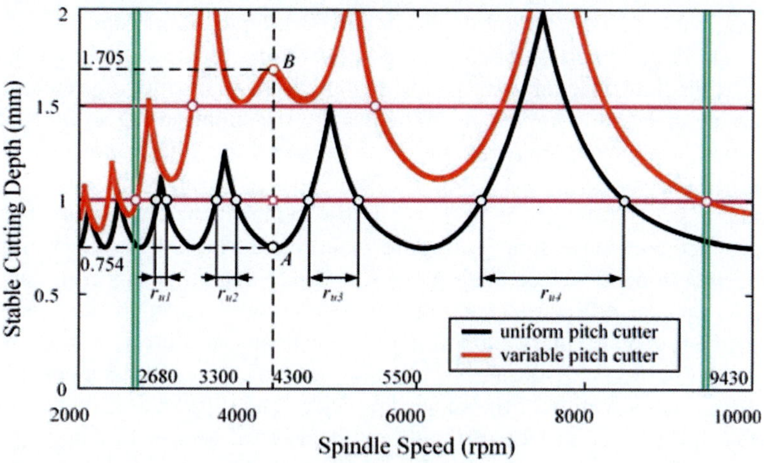

Fig. 5.16 Stability analysis of uniform and variable pitch cutters

the range from 2680 to 9430 r/min. (2) when the spindle speed was 4300 r/min, for the uniform pitch cutter, the maximum stable depth of cut was 0.754 mm (see point A). And for the variable pitch cutter, the maximum stable depth of cut improved to 1.705 mm (see point B).

5.6 Control Method of Workpiece-Fixture Subsystem Dynamic Characteristics

The processing of thin-walled parts usually needs to design and employ special fixtures. On the one hand, effective clamping and precise positioning of the workpiece can be guaranteed, on the other hand, the process rigidity of the weak parts of the workpiece could be enhanced, thereby improving processing accuracy and surface quality. For aviation thin-walled structural parts, dynamic response control can be achieved by adjusting the workpiece-fixture subsystem.

5.6.1 Control Method Based on Additional Auxiliary Support

For thin-walled structural parts, conducting auxiliary support in poor rigidity areas can improve the ability to resist processing deformation and vibration. For example, in the process of large size compressor blades and annular casings of aeroengines, auxiliary support can increase the rigidity of the workpiece, thereby reducing the dynamic response under the action of alternating cutting forces [1].

Casing, with annular and thin-wall features, is a typical important part of aero-engines. This kind of part is not only the significant structures that make up engine's propulsion air flow channel, but also the installation and positioning references of load-bearing and core components. Whether the machining accuracy and surface quality meet design requirements will directly affect the safety and reliability of the engines in service. In the process of the casing, factors such as high material strength, small wall thickness, strong time-varying characteristics, and complex clamping methods cause significant deformation and vibration to the workpiece, which results in poor surface quality and low machining efficiency and makes the machining process difficult to control. This section takes casing as an example to illustrate the role of auxiliary support in dynamic response control.

As shown in Fig. 5.17, during the clamping of the casing, each support rod acts along the normal direction of the support position. The support rods do not directly contact with the inner surface of the casing because of a layer of damping material between them. Adopt support rods with elastic rubbers to control the dynamic response during the milling of the outer surface of the casing. Elastic rubbers are used to modify the local damping of the process system, which plays a role in consuming dynamic response energy. And support rods can be performed to change the local stiffness of the system and improve the rigidity of the side wall of the casing. At this time, the local dynamic equation of the process system can be expressed as:

$$m_{w,z}\ddot{z}(t) + (c_{w,z} + \Delta c_{w,z})\dot{z}(t) + (k_{w,z} + \Delta k_{w,z})z(t) = F_z(t) \tag{5.62}$$

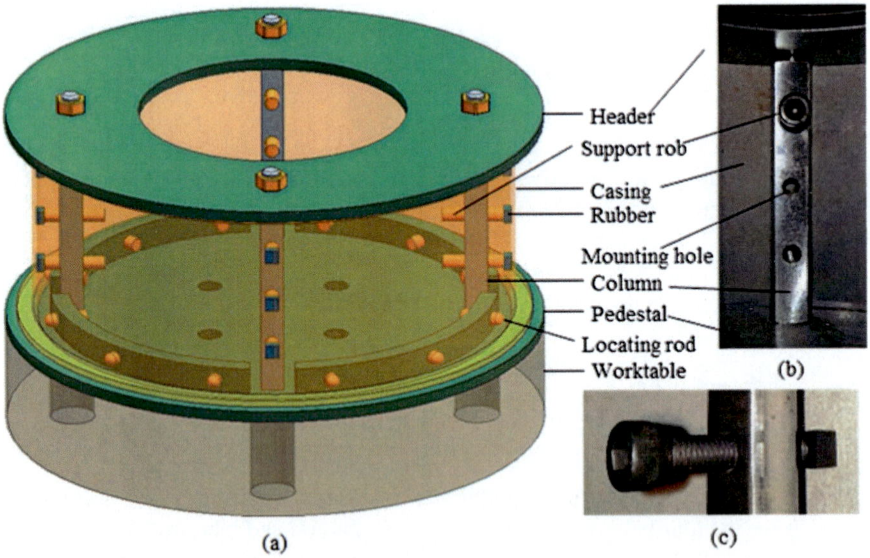

Fig. 5.17 Auxiliary support schematic diagram of the annular casing

where $m_{w,z}$, $c_{w,z}$ and $k_{w,z}$ are the normal modal mass, damping and stiffness, $\Delta c_{w,z}$ and $\Delta k_{w,z}$ are disturbances of normal modal damping and stiffness, $z(t)$ is the normal displacement, and $F_z(t)$ is the normal cutting force.

As shown in Fig. 5.18, for three different test positions in the casing, Position 1 has a support rod, Position 2 is adjacent to Position 1, and Position 3 is the middle of two support rods. Carry out milling tests and monitor cutting force at Positions 1, 2, and 3. The measurement results are shown in Fig. 5.19. The cutting parameters are as follows: spindle speed is 5000 r/min, feed speed is 320 mm/min, lead angle is 4°, radial cutting width is 12 mm, and axial cutting depth is 0.5 mm. It can be found that the cutting force at Position 1 is very stable, and the amplitude corresponding to the tooth pass frequency, i.e., 333.3 Hz, is the largest, indicating that support rods efficiently suppress the vibration of Position 1. The cutting force at Position 2, which is close to Position 1, is also relatively stable. The amplitude of the tooth passing frequency, that is, 333.3 Hz, is larger than that of any other frequencies, demonstrating the support rods can improve the local dynamic characteristics near Position 1. However, the cutting force at Position 3 fluctuates greatly, and the amplitude corresponding to other frequencies such as 119.0 and 238.5 Hz are larger than that of the tooth pass frequency, that is, 333.3 Hz, which means the milling process has obvious cutting vibration.

From the above analysis, it can be seen that using support rods with elastic rubbers to optimize local clamping structures of the casing can significantly improve the local dynamics of the process system. Compared with other active dynamic response control methods, this approach only needs to modify the fixture locally, which is convenient for implementation in the workshop. At the same time, this method can provide solutions for static problems such as tool deformation.

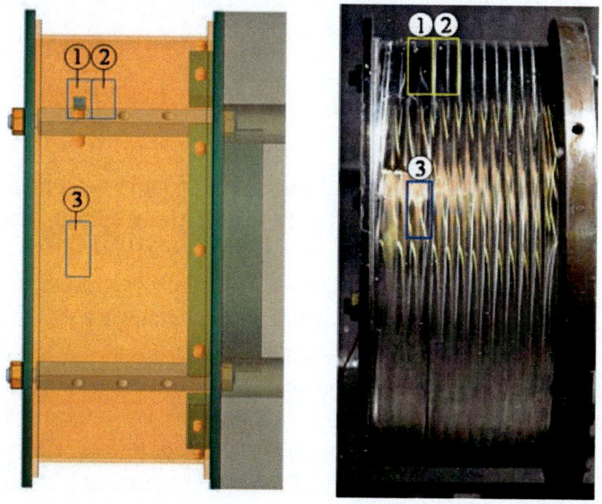

Fig. 5.18 Test positions of auxiliary supports

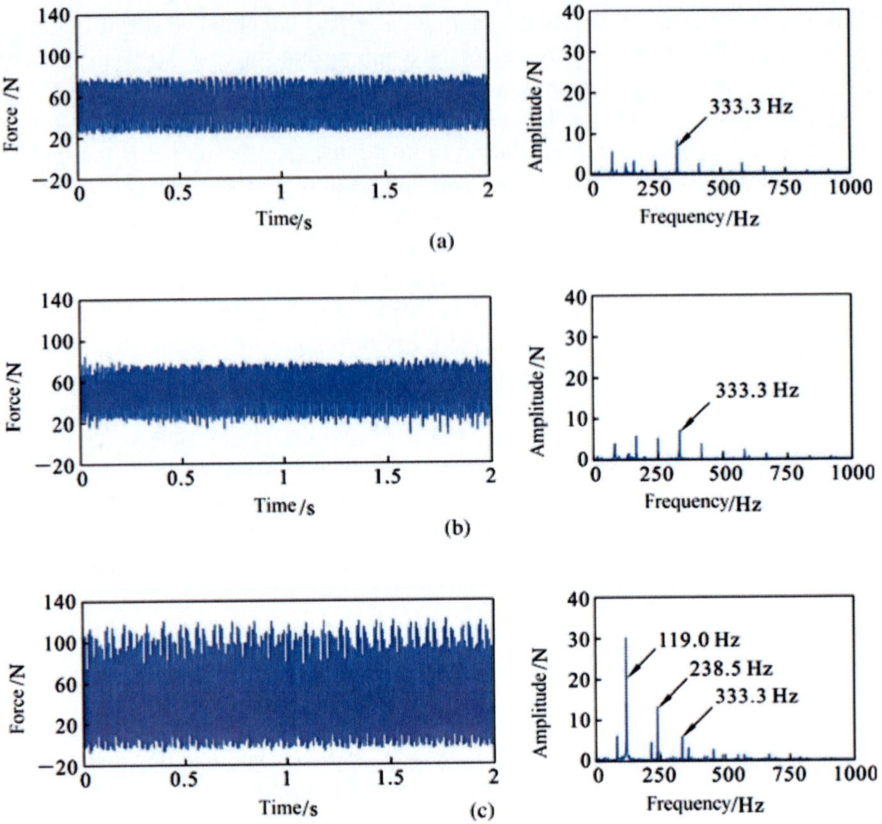

Fig. 5.19 Cutting forces and frequency spectrum of different test positions in z direction. **a** Position 1. **b** Position 2. **c** Position 3

5.6.2 Control Method Based on Additional Masses

For thin-walled structural parts, adding vibration absorbing devices to low rigidity areas can improve the ability to resist machining vibration. Taking the processing of thin-walled casing as an example, the additional masses can absorb and consume the vibration energy of side walls, thereby suppressing the vibration. The additional masses do not directly contact the inner surface of the casing, and there is a layer of elastic rubbers between them. Use additional masses with elastic rubbers to control the dynamic response when milling the outer surface of the casing. The elastic rubbers are adopted to modify the local damping of the process system, and the additional masses are performed to change the local mass. Both can play a role in consuming dynamic response energy. The local dynamic equation of the process system can be expressed as [1]:

$$(m_{w,z} + \Delta m_{w,z})\ddot{z}(t) + (c_{w,z} + \Delta c_{w,z})\dot{z}(t) + k_{w,z}z(t) = F_z(t) \tag{5.63}$$

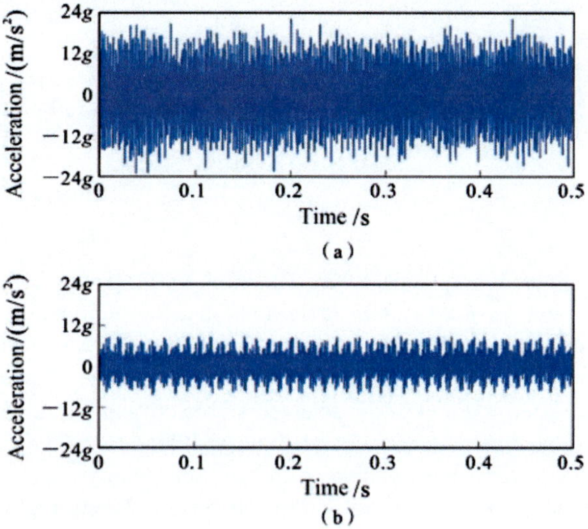

Fig. 5.20 Accelerations in the machined surface of the casing with/without additional masses. **a** Acceleration without additional masses. **b** Acceleration with additional masses

where $\Delta m_{w,z}$ is the disturbance of normal modal mass.

The milling test was carried out under the conditions of no additional masses and with an additional mass of 20 g. The accelerations at corresponding positions are collected as shown in Fig. 5.20. Milling test parameters are as follows: spindle speed is 5000 r/min, feed speed is 320 mm/min, radial cutting width is 12 mm, and axial cutting depth is 0.5 mm. It can be found that when there are no additional masses, the amplitude of accelerations is large, that is, −20g–20 g. When attaching 20 g, the amplitude of acceleration is obviously suppressed, i.e., −8g–8 g.

From the above analysis, it can be seen that adding additional masses with elastic rubbers in the inner wall of the casing can effectively reduce the vibration amplitude of the machined surface, thereby improving the machining quality of the casing. This approach can cooperate with the auxiliary support method. The support rods are installed on the columns to provide normal support and the additional masses can be flexibly attached to the inner wall of the casing with low rigidity and high deformation and vibration. The combination of the two methods can more effectively suppress the dynamic response during processing.

5.6.3 Control Method Based on Magnetorheological Damping Support

In the machining process of thin-walled parts, the removal of workpiece material has a significant impact on the stability, natural frequency, process stiffness and other dynamic parameters of the workpiece-fixture subsystem. In fact, the material removal process of the workpiece is the evolution of the dynamic characteristics of the

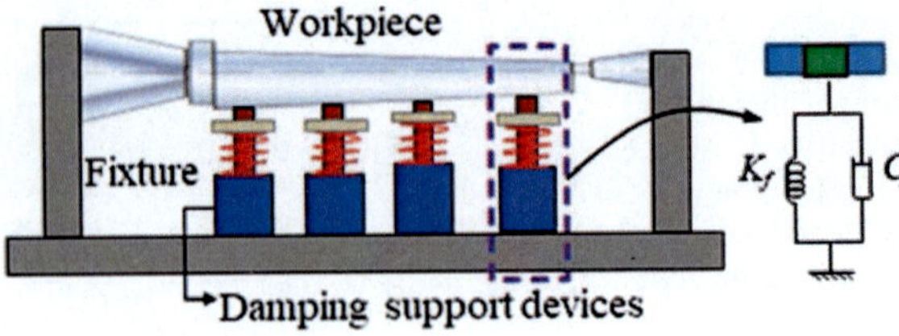

Fig. 5.21 Schematic diagram of reconstructed magnetorheological damping support for the fixture-workpiece system

fixture-workpiece system. This time-varying character makes the stable cutting state slowly evolve to an unstable state. In different machining stages, variations in system dynamic parameters decrease the stability of the system. The main performance is that the limit cutting depth is reduced and the chatter SLD is offset which causes chatter and makes the selected stable cutting parameters unstable as the machining progresses. All mentioned above will lead to difficult control on surface quality.

Aiming at the problem of chatter caused by the evolution of the dynamic characteristics of the fixture-workpiece system during the machining process, in order to improve the deficiency of chatter suppression through methods such as machining parameters optimization and machining system stiffness enhancement, magnetorheological damping support with a controllable magnetic field can be adopted to realize the reconstruction of the dynamic characteristics of the fixture-workpiece system and ensure the stability of the machining system. The reconstructed magnetorheological damping support for a fixture-workpiece system which mainly includes initial positioning fixture, thin-walled part, damper, external power supply device, etc. is shown in Fig. 5.21 [4].

Utilizing magnetorheological damping support on the thin-walled parts can realize the control of the vibration response in milling process. According to the actual conditions in the machining process, the current of the magnetorheological damping support can be adjusted, and the dynamic damping and stiffness of the damping support could be changed to achieve the reconstruction of the dynamic characteristics of the fixture-workpiece system and ensure the stable machining of the workpiece. The damping characteristics of the magnetorheological damping support are used to modify the local damping of the fixture-workpiece system, and play a role in consuming vibration energy. And its stiffness characteristics are operated to modify the local stiffness to improve the rigidity of the thin-walled workpiece. Therefore, when the magnetorheological damping support acts on the thin-walled part, the dynamics equation of the fixture-workpiece system after the local modification can be expressed as [1]:

$$\boldsymbol{M}\ddot{\boldsymbol{q}} + (\boldsymbol{C} + \boldsymbol{C}_f)\dot{\boldsymbol{q}} + (\boldsymbol{K} + \boldsymbol{K}_f)\boldsymbol{q} + P_z z = \boldsymbol{F}_z(t) \tag{5.64}$$

where $\boldsymbol{C}_f$ is the normal damping matrix, and $\boldsymbol{K}_f$ is the normal stiffness matrix.

In the actual machinig process, with the material being removed, the dynamic characteristics of the fixture-workpiece system are constantly changing, which will cause instability for the processing system. Therefore, the dynamic characteristics of the magnetorheological damping support for the fixture-workpiece system can be reasonably adjusted according to the actual working conditions to realize the reconstruction of the dynamic characteristics of the process system and maintain a stable cutting state for a long time. In this case, the dynamics equation can be expressed as:

$$(\boldsymbol{M} + \Delta\boldsymbol{M})\ddot{\boldsymbol{q}} + (\boldsymbol{C} + \Delta\boldsymbol{C} + \boldsymbol{C}_f)\dot{\boldsymbol{q}} + \left(\boldsymbol{K} + \Delta\boldsymbol{K} + \boldsymbol{K}_f\right)\boldsymbol{q} + P_z z = \boldsymbol{F}_z(t) \quad (5.65)$$

where $\Delta\boldsymbol{M}$, $\Delta\boldsymbol{C}$, and $\Delta\boldsymbol{K}$ is the variation of modal mass, damping and stiffness because of the removal of materials.

In the magnetorheological damping support for the fixture-workpiece system, by adjusting the current control parameters of the magnetorheological damping support, the dynamic output characteristics of the damping support, as well as the stiffness and damping performance of the fixture-workpiece system can be changed, so as to achieve the purpose of improving the dynamic characteristics of magnetorheological damping support for the fixture-workpiece system. Based on the modal tests, under different current control parameters, the SLD and system parameters of the magnetorheological damping support for the fixture-workpiece system are shown in Fig. 5.22 and Table 5.2, respectively.

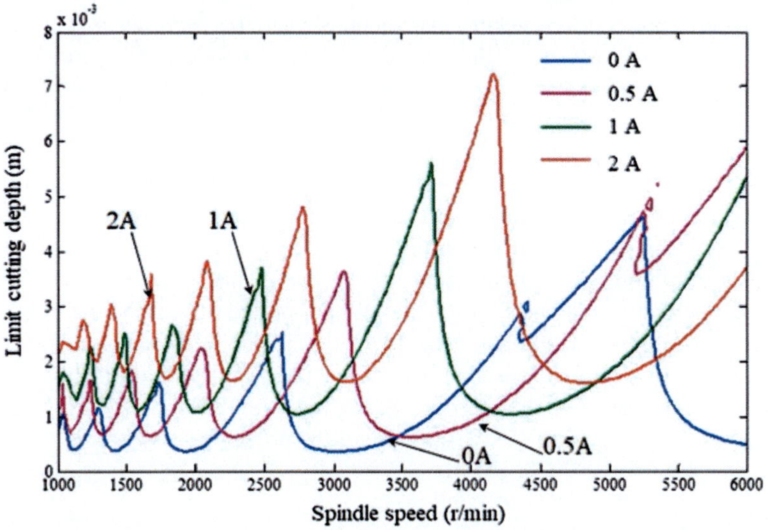

Fig. 5.22 The SLD of magnetorheological damping support for a fixture-workpiece system under different currents

Table. 5.2 Parameters of magnetorheological damping support for the fixture-workpiece system under different currents

Dynamic parameter	Current (A)			
	0	0.5	1	2
First order natural frequency (Hz)	172.86	203.6	229.8	273.483
Damping ratio	0.025	0.031	0.0358	0.0433
Normal stiffness ($\times 10^6$ N/m)	1.1930	1.3707	1.7612	2.3781

It can be seen from Fig. 5.22 and Table 5.2 that as the current of the magnetorheological damping support for the fixture-workpiece system increases, the support force and normal stiffness of the thin-walled workpiece will rise too. This situation could cause a significant increase in the natural frequency of the magnetorheological damping support for the fixture-workpiece system, and let SLD move horizontally to the right, thereby altering the machining stability under certain parameters. Thus, the magnetorheological damping support changes the stiffness characteristics of thin-walled parts and can effectively improve the dynamic stability of the milling system. However, since the magnetorheological fluid smart material in the magnetorheological damping support has magnetic saturation characteristics, its natural frequency must be adjusted within a certain range, the vibration cannot be suppressed by increasing the natural frequency of the system without restriction.

Based on the SLD shown in Fig. 5.22, select parameters at a spindle speed of 2000 r/min, feed speed of 320 mm/min, axial cutting depth of 1 mm, tool diameter of 12 mm, and teeth number of 2. Using constant cutting parameters, the milling test was completed under different currents, which verified that the control of current can change the dynamic response characteristics of the system. The measured vibration response and machined surface characteristics during the milling process are shown in Fig. 5.23.

It can be seen from Fig. 5.23 that when there is no current, the vibration acceleration has a large amplitude and severe fluctuation, and the machined surface is rough and has vibration waviness, which indicates that the machining process was in chatter. When the current is 0.5A, the vibration amplitude is significantly reduced, but the fluctuation is still obvious. Although the surface quality has been improved, there are still existing vibration waviness and chatter. When applied currents are 1A and 2A, the acceleration amplitude is $\pm 20g$ and the fluctuation is stable. The machined surface is smooth and there is no chatter, which verifies that reasonably controlling the dynamic characteristics of the magnetorheological damping support for the fixture-workpiece system can effectively improve the stability of the processing system.

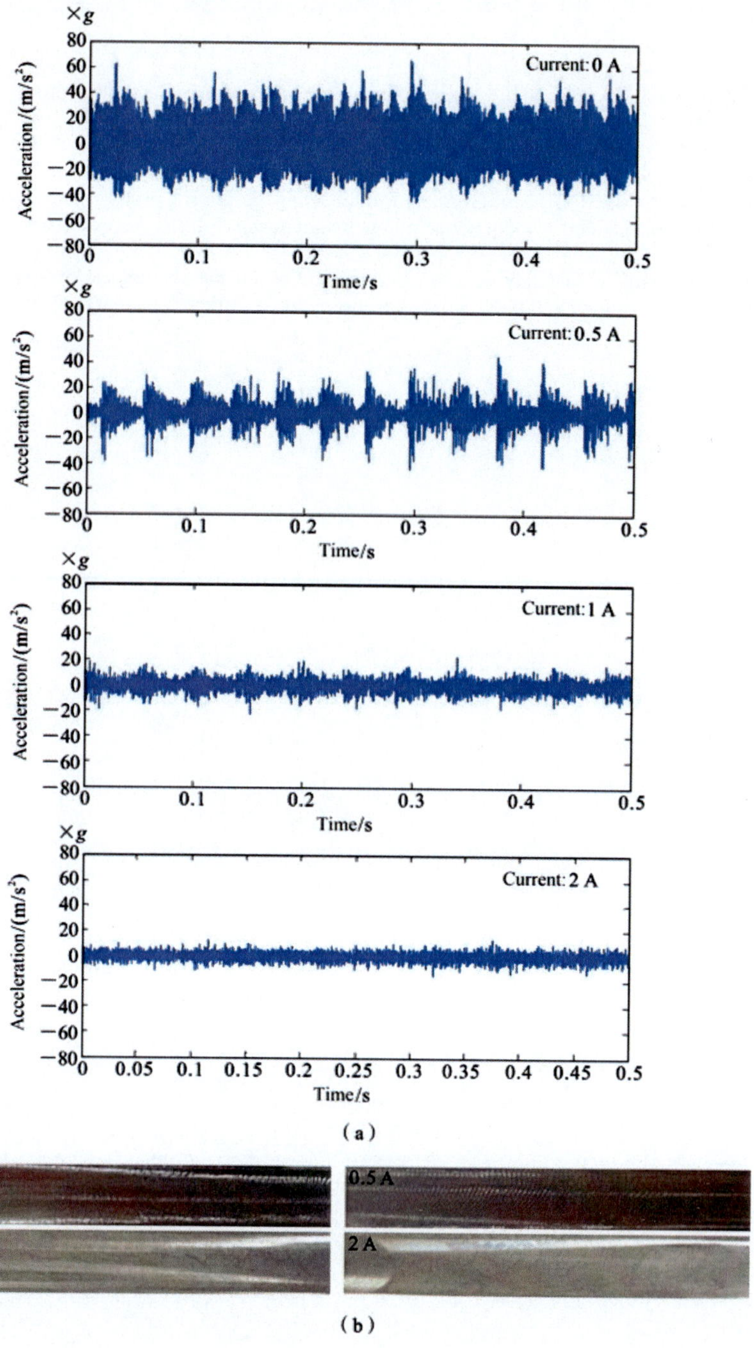

Fig. 5.23 Vibration response and machined surface characteristics of thin-wall parts in milling **a** vibration acceleration response **b** machined surface characteristics

References

1. ZHOU X. Research on dynamic response prediction and control in the milling process of ring-shaped [D]. Xi'an; Northwestern Polytechnical University, 2017.
2. ALTINTAS Y. Manufacturing automation: metal cutting mechanics, machine tool vibration, and CNC design [M]. 2nd. Cambridge, UK: Cambridge University Press, 2012.
3. MEI J W, LUO M, GUO J L, et al. Analytical modeling, design and performance evaluation of chatter-free milling cutter with alternating pitch variations [J]. IEEE Access, 2018, 6: 32367–32375.
4. MA J J. Dynamic model and reconstruction method of magnetorheological damping support fixture-workpiece system [D]. Xi'an; Northwestern Polytechnical University, 2017.

Chapter 6
Clamping Perception for Residual Stress-Induced Deformation of Thin-Walled Parts

In practical applications, thin-walled parts require very precise assembly and positioning references, which means that high manufacturing accuracy is required. CNC machining is the main manufacturing technology for thin-walled aviation parts, and it is one of the key factors that determine the quality and efficiency of machining. Thin-walled parts have a complex structure, thin walls, low rigidity and uneven distribution, making it very difficult to machining. In CNC machining, cutting force and thermal load, clamping force load and machining induced residual stress (MIRS), etc., act on the workpiece, causing deformation of the workpiece, affecting the control of machining accuracy and the improvement of machining efficiency. At present, the solution adopted in production is to use conservative cutting parameters, and perform secondary compensation processing or manual calibration after the processing is completed. This has caused a reduction in production quality and efficiency.

The MIRS-induced deformation introduced by CNC machining is an important reason for the out-of-tolerance or scrap of parts. When cutting metal materials, the initial residual stress (IRS) state of the blank material is broken; at the same time, the material in the contact part of the workpiece and the tool undergoes plastic deformation during the machining process, and considerable MIRS is generated in the thin layer under the workpiece surface. Because thin-walled parts have the characteristics of a thin wall and low rigidity, after the clamping is released, the residual stress rebound deformation is very easy to occur, resulting in the overall deformation of the part. How to predict and evaluate the residual stress-induced deformation (RSID) caused by the numerical control and machining of thin-walled parts and optimize the control of the process is a hot topic of current research, and it is also a huge challenge for the improvement of the processing quality of the country's aviation complex thin-walled parts [1, 2].

D. Zhang et al., *Intelligent Machining of Complex Aviation Components*, Research on Intelligent Manufacturing, https://doi.org/10.1007/978-981-16-1586-3_6

6.1 Residual Stress in Cutting Process

The cutting process of metal is essentially a process in which metal is squeezed and deformed and finally yielded and fractured. When the metal is squeezed, the principal stress is generated inside it, and the maximum shear stress is also generated in the direction at a 45° angle to the extrusion direction. When the shear stress reaches the yield limit of the metal material, the material will yield and produce shear slips and eventually break. The metal cutting process can be regarded as the cyclic extrusion process of the tool to the workpiece material. The tool continuously squeezes the workpiece material and causes it to slip and fracture in the direction of the maximum shear stress to achieve material removal.

Figure 6.1 shows the cutting process and the mechanism of residual stress generation. During cutting, the workpiece material at the front end of the edge of the tool tip will flow in two directions. The material above the O point will flow out from the rake face and eventually form chips, and the part below the O point will be squeezed by the processing tool and stay on the processed surface to form a processed surface. The plastic deformation of this layer of material left on the processed surface is the main cause of residual stress. At the front end of the tool edge, the workpiece material is subjected to the cutting shear force, and the material crystal grains will appear plastic shrinkage along the cutting direction, and plastic stretch will occur perpendicular to the cutting direction. This phenomenon is called the "shaping protrusion effect," and the shaping protrusion effect will cause residual tensile stress on the processed surface. As the tool continues to cut, the processed surface material will be affected by the ploughing force and friction of the tool flank. The tool will squeeze and rub the surface material, and the surface material will undergo further deformation. This phenomenon is called the "squeezing effect". The squeezing effect often creates residual compressive stress on the processed surface. These two effects are mainly produced by the action of cutting force, collectively referred to as mechanical effects.

During the metal cutting process, violent shear, extrusion, friction and material yield flow will occur at the tool-workpiece contact interface. The cutting force will generate a lot of cutting heat, and the temperature of the tool-workpiece contact area

Fig. 6.1 Cutting process and residual stress generation mechanism

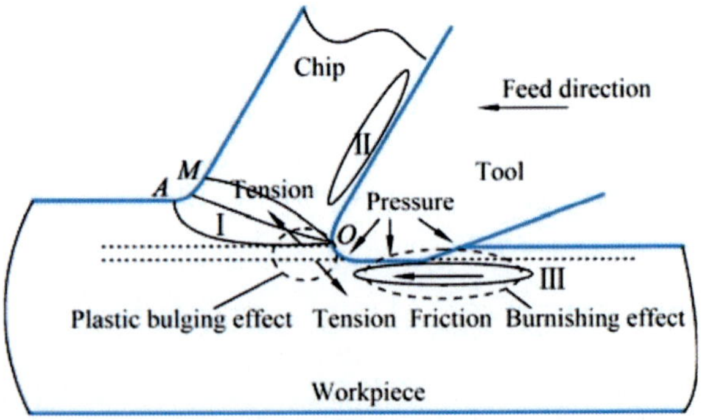

will gradually rise. In normal cutting, the temperature in the cutting area can reach several hundred degrees Celsius, and the cutting temperature field has a very large change gradient. Due to the existence of a large temperature gradient, the temperature drops sharply from the surface of the workpiece to the inside of the workpiece. During cutting, the surface material of the workpiece will thermally expand and cause tensile deformation. At the same time, the surface material is constrained by the internal undeformed material, which causes the material to suffer pressure. After the cutting process is completed, the temperature of the workpiece surface and internal materials will drop to room temperature. At this time, the workpiece material shrinks, the workpiece surface shrinks more, but the internal shrinkage is less. The shrinkage of the surface material is constrained by the inner material, and finally, there is the residue on the surface of the workpiece. Tensile stress, residual compressive stress appears in the inner layer. Therefore, the higher the temperature field amplitude and temperature gradient of the machined surface, the greater the residual tensile stress on the surface of the workpiece. The influence caused by the uneven cutting temperature distribution inside the workpiece is collectively called the thermal effect.

During the cutting process, thermal and mechanical effects are produced at the same time, and corresponding thermal and mechanical stresses are generated on the surface of the workpiece. The superposition of the two determines the residual stress distribution on the surface of the workpiece after cutting. In other words, the thermal coupling effect is the decisive factor for the residual stress distribution on the surface of the workpiece after cutting. The residual stress distribution on the surface of the workpiece caused by cutting is characterized as:

$$\sigma_T \approx \sigma_{\text{Mech}} + \sigma_{\text{Therm}} \tag{6.1}$$

where: σ_T is the residual stress on the surface of the workpiece caused by cutting; σ_{Mech} is the residual stress introduced by the mechanical effect; σ_{Therm} is the residual stress introduced by the thermal effect.

The residual stress distribution of the milling surface is caused by the thermal coupling effect on the contact interface, and the thermal coupling state of the cutting process is closely related to the cutting conditions. According to the different cutting conditions of the workpiece surface, the residual stress on the surface of the workpiece will show different distribution forms.

6.2 Residual Stress-Induced Deformation

According to different deformation mechanisms, the RSID of thin-walled parts can be divided into two types: IRS-induced deformation and MIRS-induced deformation. On the one hand, the IRS will inevitably be introduced in the manufacturing process of the blank material. Although most of the blanks of aerospace parts have undergone stress-relief treatment, the IRS cannot be completely eliminated. After the stress-relief treatment, the residual stress of the blanks is mostly around tens of MPa. During the machining process, especially during the rough machining process, a large amount of material is removed, and the internal stress balance of the blank is broken, resulting in redistribution and rebalance of internal stress. In this process, the stress rebalance produces strain, which results in IRS-induced deformation. On the other hand, when the material is removed, there is a strong thermal coupling effect on the tool-workpiece contact interface, and the surface of the workpiece is plastically deformed, which cannot be recovered after the thermal load is unloaded, resulting in large residual stress on the surface of the machined area. This stress varies greatly along the depth direction on the surface, and the range is within a few hundred microns of the surface. The introduction of MIRS produces an equal moment of force, which causes the RSID of the workpiece. In general, the two forms of deformation work together to determine the final deformation state of the part. When the wall thickness of the workpiece is large, the rigidity of the workpiece is large, and the residual stress acting layer is very thin, so its contribution to the deformation of the part is small. However, in the machining of thin-walled parts, the final thickness of the part may reach 1 mm or less, and the rigidity of the workpiece is very small. At this time, the processing residual stress layer has accounted for a considerable proportion of the thickness of the entire workpiece, and the deformation due to the residual stress is very significant.

As shown in Fig. 6.2, assuming that the initial state of the blank is in the stress balance state, the part has not deformed, as shown in the state *a* in Fig. 6.2. The IRS-induced deformation is caused by the removal of the blank material, which is equivalent to the removal of part of the stress spring (residual stress) due to the removal of the material. The internal stress balance of the part changes and the part deforms to restore the balanced state, resulting in the part's failure. The IRS is deformed, as shown in state *b* in Fig. 6.2. Correspondingly, cutting processing will introduce MIRS on the surface of the workpiece, which is equivalent to adding a stress spring to the original workpiece system, which will also cause the imbalance of internal stress and cause deformation of the workpiece, as shown in the state *c* in Fig. 6.2, which is the MIRS induced deformation.

In the entire machining process of the part, these two kinds of RSID exist at the same time. According to the principle of deformation superposition, the superposition of the two shows the final RSID of the part. According to the generation mechanism of these two kinds of deformation, the machining process can be divided into rough machining, semi-finishing and finishing processes for analysis. As shown in Fig. 6.3, for thin-walled parts with difficult-to-machine materials in aeroengines, in the rough machining stage, the parts are thicker and have high rigidity, and a large amount of rough material is removed. The IRS-induced deformation is closely related to the amount of material removal. The IRS-induced deformation is relatively large and occupies an absolute dominant position. It can be approximated that the IRS-induced deformation is the RSID in this process. In the semi-finishing and finishing stages, the amount of material removal is small. At this time, the part will be processed to the final part thickness, with a small wall thickness. At this time, the main effect on

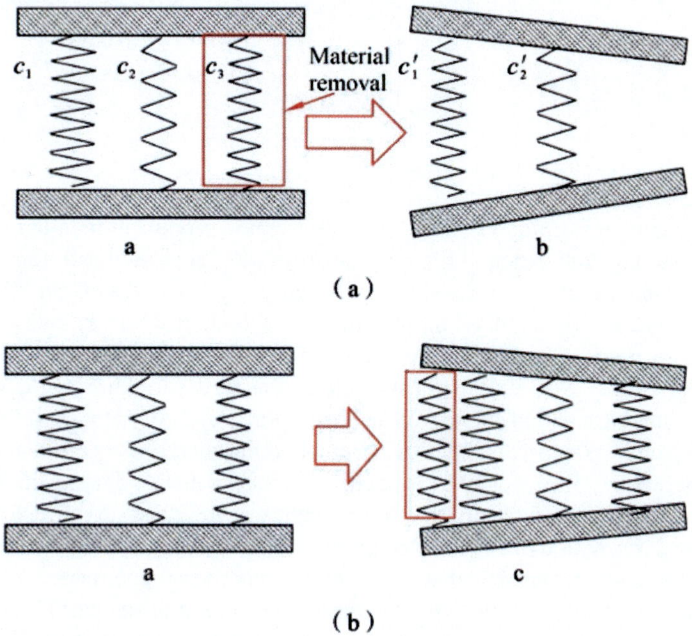

Fig. 6.2 Spring model of residual stress. **a** IRS-induced deformation **b** MIRS-induced deformation

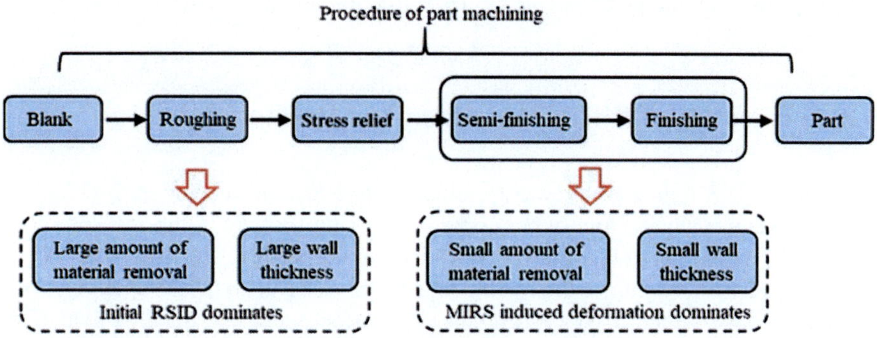

Fig. 6.3 Analysis of RSID in each process state

the deformation is the surface residual stress introduced by cutting, and the residual stress and deformation of the machining will play a leading role.

According to the principle of material mechanics, when a rectangular coordinate system is established on a machined part, the internal stress at any point inside it can be expressed as the following tensor form [3]:

$$\sigma = \begin{bmatrix} \sigma_{xx} & \tau_{xy} & \tau_{xz} \\ \tau_{yx} & \sigma_{yy} & \tau_{yz} \\ \tau_{zx} & \tau_{zy} & \sigma_{zz} \end{bmatrix} \tag{6.2}$$

where $\sigma_{xx}, \sigma_{yy}, \sigma_{zz}$ are normal stress components; $\tau_{xy}, \tau_{yx}, \tau_{xz}, \tau_{zx}, \tau_{yz}, \tau_{zy}$ are shear stress components.

The normal stress component and the shear stress component together constitute a stress state. The unbalance of the normal stress will cause the bending of the part, and the unbalance of the shear stress will cause the shear distortion of the part. Among the three normal stress components, σ_{zz} is the normal stress in the thickness direction. For thin-walled parts, the bending stiffness in the thickness direction is the largest, and σ_{zz} has the smallest contribution to deformation. In addition, in actual production, most of the blanks used are pre-stretched plates or rolled plates, which have undergone stress relief treatment, and the shear force component is very small. Therefore, in the current research, researchers mostly ignore the influence of the normal stress and shear stress components in the thickness direction on the processing deformation of thin-walled parts, and focus on analyzing the two normal stress components σ_{xx} and σ_{yy} that can be effectively measured now. σ_{xx} and σ_{yy} are plane stresses, corresponding to the bending deformation of the part along the XX direction and the YY direction, respectively. Based on the above reasons, for the IRS field, the influence of the normal stress in the length and width directions of the part on the IRS induced deformation can be analyzed; for the MIRS field, the analysis can be focused on the positive cutting feed direction and the vertical feed direction.

6.3 Principles of RSID Perception and Prediction

The current IRS field measurement method is to slice and discretize the blank to obtain the stress state. There is an assumption of uniformity in the measurement. In actual processing, for some parts with complex shapes, the IRS field is not uniformly distributed during the manufacturing process due to the complexity of the blank boundary, which makes it difficult to accurately measure the IRS field. This makes it difficult to accurately obtain the IRS state of the part blank, so the RSID prediction method based on the stress field is not applicable. In addition, during the milling process, the working condition factors have an important influence on the processing quality, which will directly determine the final state of residual stress in the surface layer, thereby affecting the generation of residual stress and deformation. The existing processing residual stress prediction method is based on working condition mapping and stress fitting, and the machining process is required to be discrete into steady working conditions. However, due to the time-varying working conditions, the residual stress field is not uniformly distributed on the surface of the workpiece, and it is difficult to evaluate the surface stress field by measurement. The RSID prediction method based on the perception of deformation potential energy

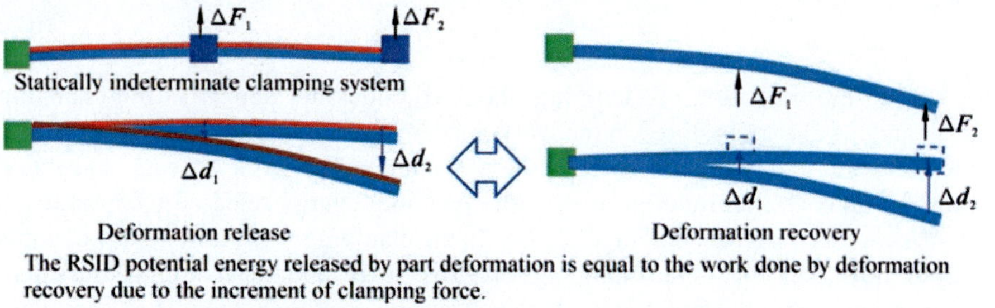

Fig. 6.4 Principles of RSID perception and prediction

is proposed to address this problem. The so-called perception prediction of RSID means that in the actual machining process, the monitorable quantity that changes due to the introduction of residual stress is monitored through the sensor to obtain the change of the monitored quantity before and after processing. On this basis, the residual deformation potential energy of the workpiece in the clamping state is sensed, the mapping relationship between the monitoring quantity and the final RSID is established, and the RSID in the free state after the clamping and unloading of the part is predicted. The principle is shown in Fig. 6.4.

The following describes the principle of perception and prediction of RSID from the perspective of energy. Material removal introduces an additional stress field inside the workpiece. This additional stress field is composed of the additional stress field introduced by the rebalance of the IRS field and the processing residual stress field caused by thermal and mechanical coupling. Assuming that the additional stress field is known and only the normal stress is considered, the total energy introduced into the workpiece can be expressed as:

$$E_0 = \frac{1}{2} \iiint_V \left(\sigma_{xx}\varepsilon_x + \sigma_{yy}\varepsilon_y\right) dxdydz \tag{6.3}$$

where E_0 is the energy introduced by cutting; σ_{xx} and σ_{yy} are stress; ε_x and ε_y are strain.

After the clamping is released, the workpiece undergoes RSID under the action of the residual stress and other moments. Define the residual energy in the workpiece after deformation as E_1, and E_1 is the potential energy inside the part at this time. The RSID work meets the principle of minimum potential energy, that is, when the equilibrium state is reached, in the displacement field meeting the given boundary, the real displacement field minimizes the potential energy of the workpiece [4], namely:

$$\frac{\partial E_1}{\partial u_i} = 0 \tag{6.4}$$

Define E_d as the energy released by the deformation of the workpiece caused by the unbalanced residual stress after the clamping is released, it can be expressed as:

$$E_d = E_0 - E_1 \tag{6.5}$$

Due to the limitation of clamping, the RSID does not appear in the clamping state, and the energy is stored in the workpiece in the form of potential energy. After clamping and unloading, this part of the potential energy work appears as residual stress springback deformation, that is, the potential energy released by springback deformation is the RSID potential energy. In the clamping state, the existence of this part of the energy causes the clamping force to change, and the clamping stress field balances the additional stress field caused by cutting. As shown in Fig. 6.4, according to the principle of virtual work, after the clamping point is released, the workpiece undergoes RSID. The RSID potential energy released by the workpiece is equal to the work done by the reverse of the clamping force increment to restore the deformation. At the same time, the displacement of the clamping point of the workpiece can also be approximately expressed as a function of the change in clamping force, namely:

$$E_d = E_0 - E_1 = \sum_{i=1}^{n} \Delta F_i \Delta d_i = \sum_{i=1}^{n} \Delta F_i f(\Delta F_i) \tag{6.6}$$

where ΔF_i is the change value of the clamping force at the i-th clamping position; Δd_i is the deformation after the i-th clamping position is released.

According to Eq. (6.6), there is a mapping relationship between the change of clamping force and the RSID potential energy. By monitoring the change of the clamping force of the workpiece, the overall residual deformation potential energy can be evaluated, and the purpose of predicting the RSID of the part can be achieved.

6.4 RSID Perception Prediction Model

In order to realize the perception and prediction of RSID, a corresponding mathematical model needs to be established. Note that D is the perception target matrix (predicted RSID of the part), $\Delta \boldsymbol{F}$ is the perception matrix (change value of clamping force), f is the mapping relationship from the perception matrix to the target matrix, then the general expression of RSID perception prediction model can be expressed as:

$$\boldsymbol{D} = f(\Delta \boldsymbol{F}) \tag{6.7}$$

In the equation, its components are:

$$\boldsymbol{D} = \begin{bmatrix} \boldsymbol{D}_1 & \boldsymbol{D}_2 & \cdots & \boldsymbol{D}_m \end{bmatrix}^T \tag{6.8}$$

$$\Delta F = \left[\Delta F_1 \ \Delta F_2 \ \cdots \ \Delta F_m \right]^T \tag{6.9}$$

The perception matrix ΔF represents the change in clamping force at different perception moments (1, 2, ..., m) at different perception positions (1, 2, ..., n). Define the initial clamping force as $F_0 = [F_{01} F_{02} \cdots F_{0n}]$, each element in the vector represents the clamping force at the initial moment 0 sensing position $(1, 2, \ldots, \text{n})$, and the value of each element needs to be determined by a pressure sensor. The machining process of the part is equivalent to providing excitation to the RSID sensing system, introducing the RSID potential energy to change the clamping force. The clamping force at time 1 is defined as $F_1 = [F_{11} F_{12} \cdots F_{1n}]$, then the clamping force changes Vector $\Delta F_1 = F_1 - F_0$. As the cutting process progresses, new perception moments are constantly set. The clamping forces sensed at the new moments are F_2, F_3, ..., F_m, and the corresponding expanded clamping force change vectors are ΔF_2, ΔF_3, ..., ΔF_m.

The target matrix D represents the RSID matrix of the parts at different sensing moments $(1, 2, \ldots, \text{m})$ and different sensing positions $(1, 2, \ldots, \text{k})$. Each element in the target matrix represents the residual stress and deformation of the parts at different perception and prediction moments, that is, the residual stress and deformation of the two parts at different perception and prediction positions when the clamping is released at different perception moments. Among them $D_i = [D_{i1} D_{i2} \cdots D_{ik}]$, which represents the residual stress and deformation value at different positions of the part at the i-th sensing moment. It should be noted that the value position of each element does not necessarily correspond to the clamping force-sensing position.

The most important thing to realize the perception system is to obtain the mapping relationship f between the perception matrix and the target matrix. According to the mapping relationship, the target matrix can be solved by the perception matrix, so as to realize the perception prediction of RSID. The mapping relationship can be obtained through theoretical derivation, finite element simulation or intelligent algorithms.

In the real machining process, the material removal process actively introduced residual stress or caused an imbalance of internal residual stress. This is equivalent to applying an incentive to the entire perception system, resulting in a change in the perception matrix. If the perception process is simplified, only the change in clamping force after the processing is sensed, and the stimulation of the processing causes the appearance of the perception matrix ΔF. According to the established perception mapping relationship, that is, the relationship between the perception matrix and the target matrix $f : \Delta F \rightarrow D$, the effective prediction of the target matrix D can be completed, and the mathematical model of the perception system is established accordingly.

From the above description, we can see that the mathematical model of the perception system needs to be constructed from three aspects. The first is the definition of the perception matrix, which mainly determines the perception means, perception moment and perception position. The residual stress and deformation sensing mean to monitor the change of clamping force, mainly by installing a force sensor at a specific clamping position to monitor the change of clamping force before and after machining; the sensing time can be determined according to the needs of the machining technology; and the sensing position can be based on simulation or as a result of the machining of the test piece, determine the position of weak stiffness, or apply the clamping sensing point to determine. The second is the definition of the target matrix, mainly to determine the prediction and evaluation position of residual stress and deformation. This point is mainly determined according to the process requirements. The evaluation point is defined in the position with high accuracy requirements of the part; if the accuracy requirements are consistent, the evaluation point is defined in the easily deformable position to control the maximum RSID. The third is the mapping relationship. Due to the complexity of the part structure, it is often difficult to obtain the mapping relationship by theoretical solution. Therefore, it is necessary to obtain the mapping relationship through simulation or intelligent algorithms.

6.5 Potential Energy Perception of Residual Stress and Deformation in Typical Clamping Forms

Figure 6.5 shows two commonly used workpiece-fixture contact methods in the clamping process. Figure 6.5a shows normal force contact, and Fig. 6.5b shows normal force-two-way friction contact. After the RSID potential energy is introduced into the workpiece, the simple normal contact force is rare. Under the action of the residual stress and other moments, the clamping contact is mostly normal force-two-way friction contact. Using a bidirectional force sensor to sense the change of clamping force can obtain more accurate results. However, in practical applications, especially in the processing of thin-walled parts, the existence of redundant constraints mostly restricts the deformation and vibration of the workpiece along the wall thickness direction. The stiffness of the workpiece in the wall thickness direction is much smaller than the stiffness in other directions. The deformation introduced by the residual stress is mainly in the wall thickness direction. At this time, the change of the normal contact force is much larger than the change of the friction force. Using a unidirectional force sensor to sense the change of the normal contact force can also solve the perception prediction model more accurately.

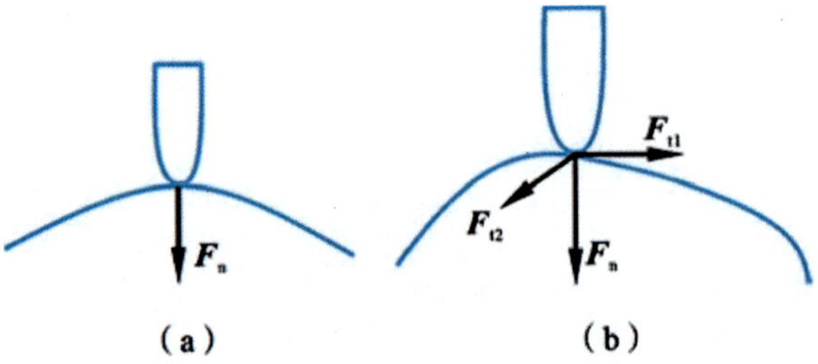

Fig. 6.5 Typical workpiece-fixture contact form. **a** normal contact force **b** normal force-two-way contact

The following analysis ignores the change in clamping contact tangential force, and focuses on the change in normal elastic contact force. The unidirectional pressure sensor can be used to monitor the change of clamping force, and then predict the RSID of the workpiece.

6.5.1 Surface Constraints in Redundant Constraints

As shown in Fig. 6.6, in practical applications, there are three types of clamping with surface constraints in common redundant constraints, namely, surface support clamping, point support clamping and face-to-face clamping. According to Eq. (6.9), the solution of the residual stress and deformation sensing model needs to sense the change of clamping force at a specific position. For the clamping method with surface constraints in multiple constraints, the equivalent moment caused by the introduction of RSID potential energy is offset by the moment generated by the change in clamping force of the workpiece by the redundant constraints. Its mathematical expression is:

$$M = \sum_{i=1}^{m} \int_{A_i} \Delta p_i dA \cdot l_i + \sum_{j=1}^{n} \Delta F_j \cdot l_j \tag{6.10}$$

where M is the equivalent moment of residual stress; m is the number of surface constraints; A_i is the contact area of the surface constraint clamping; n is the number of point constraints; Δp is the contact stress increment on the contact surface of the surface constraint; ΔF is the point constraint in the contact force increment; l is the clamping force arm.

The resultant contact force value of surface restraint can be sensed by installing a pressure sensor on the fixture. However, when the machining is completed and

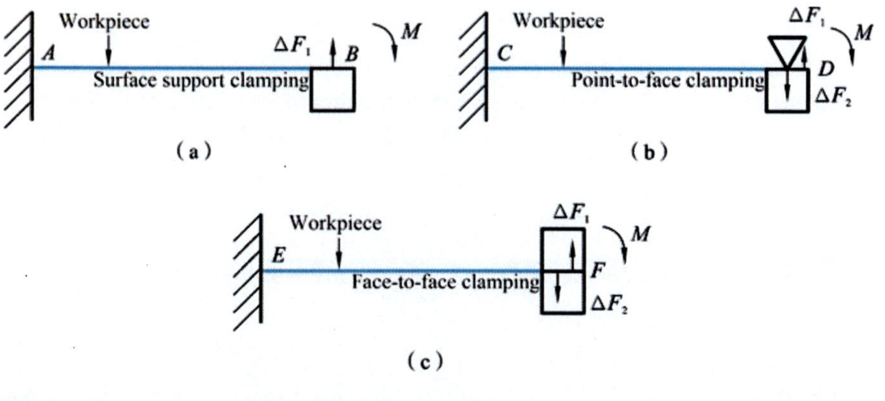

Fig. 6.6 Typical surface restraint clamping form. **a** surface support clamping form **b** point-to-face clamping form **c** face-to-face clamping form

the RSID potential energy is introduced into the workpiece, an equal force moment is introduced at one end of the workpiece. Since the surface-constrained clamping contact is a surface, the clamping force of the fixture to the workpiece is not applied directly, and it is difficult for the prior art to directly obtain an accurate resultant force position through a sensing method. The uncertainty of the position of the surface contact clamping force makes the perception model unable to be accurately solved. However, in practical applications, when the clamping surface of the fixture is small enough relative to the size of the entire workpiece, it can be approximately considered that the clamping force application position is in the middle of the clamping surface.

6.5.2 Redundant Constraints Are Point Constraints

As shown in Fig. 6.7, the redundant constraint in practical application is the point constraint. There are two clamping forms, one is the point support form, and the other is the point-to-point clamping form.

For the point-constrained clamping form, the equilibrium relationship of residual stress and moment of force is:

$$M = \sum_{i=1}^{n} \Delta F_i \cdot l_i \tag{6.11}$$

By installing a pressure sensor on a point constraint, the clamping force of a specific constraint position can be accurately sensed, that is, the elements on the right side of Eq. (6.11) can be accurately sensed. Under this type of clamping, the residual stress and deformation perception prediction model can be accurately solved. However, when applying support for deformation sensing, it is necessary to estimate the direction of deformation so that the deformation of the workpiece at the support position compresses the support.

As shown in Fig. 6.8, in the actual sensing process, due to the limitation of the clamping form and the complexity of the part structure and surface shape, it is often difficult to ensure that the sensor's sensing direction is completely consistent with the workpiece at the clamping point during the clamping process. The surface normal is consistent. At this time, the perceived clamping force is the decomposition value of

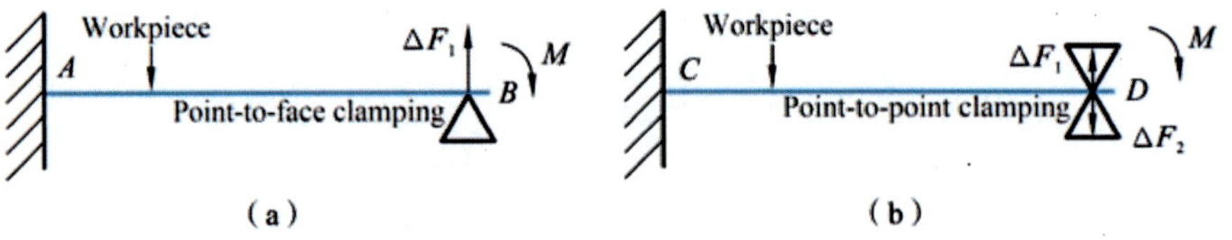

Fig. 6.7 Typical point restraint clamping system. **a** Point support clamping form **b** Point-to-point clamping form

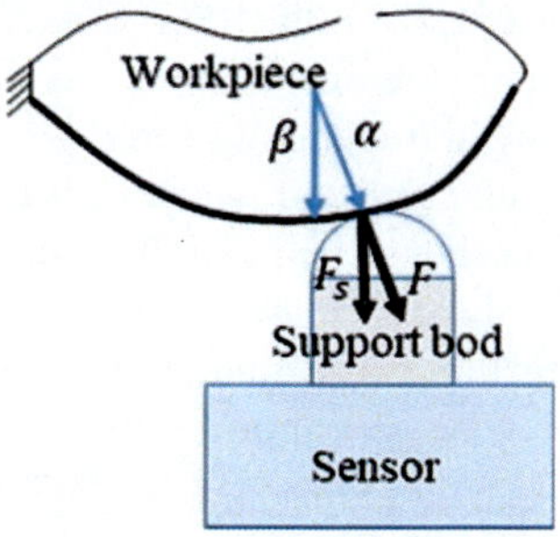

Fig. 6.8 Schematic diagram of the surface point constraint

the actual contact force in the axial direction of the sensing rod, and the contact force value can be reversed through the sensing force to solve the RSID sensing prediction model. The solution formula is as follows:

$$F = F_s \cdot \frac{|\boldsymbol{\alpha} \cdot \boldsymbol{\beta}|}{\boldsymbol{\alpha} \cdot \boldsymbol{\beta}} \tag{6.12}$$

where F is the change value of the clamping force of the sensing position; F_s is the pressure change value sensed by the sensor; $\boldsymbol{\alpha}$ and $\boldsymbol{\beta}$ are the normal vector of the curvature and the axis vector of the sensor, respectively.

6.6 Solving Residual Stress and Deformation Perception Prediction Model

6.6.1 Solution Method and Procedure

In order to ensure the stability of the clamping, the workpiece in CNC machining mostly adopts the super static clamping method. In addition to the necessary constraints to keep the geometry unchanged, there are some redundant constraints in the clamping system. The statically determinate basis of the clamping system is defined as the necessary constraints and the structure of the workpiece to maintain the geometry of the clamping system. Statically determinate basis can have different choices, not the only one. The redundant constraints of the thin-walled parts clamping system are mostly auxiliary supports to improve the rigidity of the system and reduce the vibration and deformation caused by the mechanical load during the processing.

In the statically indeterminate clamping system, according to the related theory of elastic mechanics, the application of the deformation superposition principle and deformation coordination equation can realize the solution of the RSID perception prediction model. The solution formula is:

$$D_M = -\delta \cdot \Delta F \tag{6.13}$$

where D_M is the RSID of each deformation potential energy-sensing point on the part after clamping release; δ is the deformation of each sensing point on the part under any unit clamping force. At this point, it shows that the RSID after the workpiece is clamped and unloaded is equal to the deformation caused by the reaction force of the change value of the clamping force acting on the workpiece alone.

Take the plate clamping system as an example to solve the residual stress and deformation perception prediction model of a statically indeterminate clamping system. Figure 6.9a shows a three-dimensional single-sided milling and clamping system for thin plates. The workpiece is a flat plate. One end is fixed by pressing plates or bolts, and the other end is simply supported by auxiliary supports. A pressure sensor is installed on the auxiliary support to monitor the change of the support force. There is an extra constraint in the clamping structure, which is a one-time hyperstatic clamping system.

During the milling process, the tool cutting material introduces residual compressive stress on the upper surface of the workpiece, which causes the workpiece to be convexly deformed. Simplify the workpiece to a beam, and the clamping system can be simplified to a simply supported beam structure, and its load action is shown in Fig. 6.9b. The machining process introduces residual compressive stress in the workpiece, and the effect of residual compressive stress on the deformation of the plate can be equivalent to the moment M applied to one end of the plate. If there is no redundant constraint B, the workpiece will be bent and deformed immediately. Since the excess constraint B exerts an excess constraint force ΔF_1 on the workpiece, the workpiece is not deformed in the clamping state. The energy introduced by the residual stress is stored in the workpiece in the form of deformation potential energy. Remove the excess constraint B, and replace it with the excess constraint force ΔF_1. The excess constraint force ΔF_1 is the change value of the support force sensed by the pressure sensor. Under the combined action of ΔF_1 and M, the supporting point of the end of the workpiece b moved along the direction of ΔF_1, which is defined as D_1. It can be considered that D_1 is composed of two parts: one part is the D_{1M}

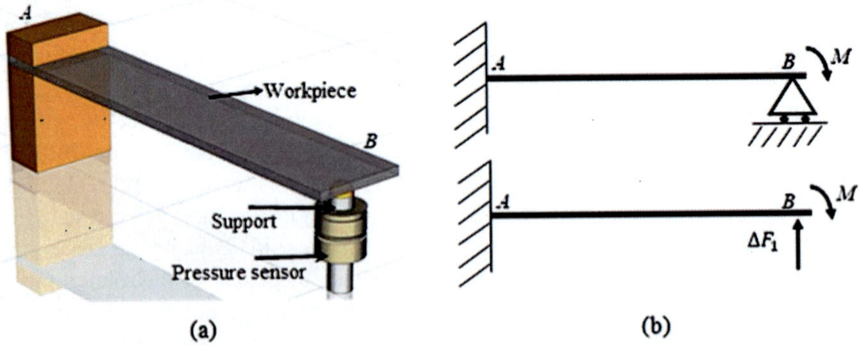

Fig. 6.9 Analysis of hyperstatic clamping system. **a** the hyperstatic clamping system **b** two-dimensional simplified force state

generated by the statically determinate base under the action of the residual stress equal moment M alone, and the other part is the displacement D_{1F} generated by the statically determinate base under the action of the excess constraint force ΔF_1, namely:

$$D_1 = D_{1M} + D_{1F} \tag{6.14}$$

where the first subscript 1 of the displacement symbol represents that the displacement occurs at the point of action F_1 at the B end and along the direction of F_1.

Due to the existence of auxiliary support B, the end B does not move, so there is:

$$D_1 = D_{1M} + D_{1F} = 0 \tag{6.15}$$

When calculating D_{1F}, a unit force can be applied along the F_1 direction on the statically determinate base, and the displacement of point B along the F_1 direction due to the unit force is recorded as δ_{11}. For linear elastic structures, the displacement is proportional to the magnitude of the force, inherently:

$$D_{1M} = -\delta_{11}\Delta F_1 \tag{6.16}$$

The left side of Eq. 6.16 is the deformation caused by the residual stress of the workpiece after the excess constraint is released, and the right side is the deformation caused by the reaction force of the supporting force change on the statically determinate foundation. In the formula, ΔF_1 can be obtained by monitoring the clamping force, and δ_{11} can be obtained by theoretical derivation or simulation. In this way, the mapping relationship between the perception quantity target matrix and the perception quantity matrix is established, and the solution of the perception prediction model of the residual stress change of a statically indeterminate clamping system is realized.

The appeal method is extended to the two-stage hyperstatic clamping system. The statically indeterminate clamping structure shown in Fig. 6.10 has two redundant constraints B and C. After the milling is completed, the force analysis is performed

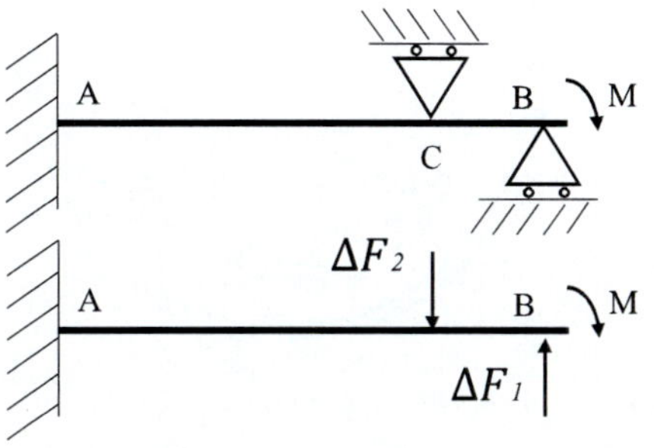

Fig. 6.10 Analysis of two-time statically indeterminate clamping system

on them. The residual stress introduced during the machining process is replaced by an equal moment of force. The redundant constraints produce two constraint inverses. Similarly, the redundant constraints B and C are released and replaced by the redundant constraints ΔF_1 and ΔF_2. Then the deformation coordination equation at the position of the redundant constraints B and C can be expressed as:

$$\begin{cases} \delta_{11}\Delta F_1 + \delta_{12}\Delta F_2 + D_{1M} = 0 \\ \delta_{21}\Delta F_1 + \delta_{22}\Delta F_2 + D_{2M} = 0 \end{cases} \tag{6.17}$$

The RSID after the clamping is removed is:

$$\begin{cases} D_{1M} = -\delta_{11}\Delta F_1 - \delta_{12}\Delta F_2 \\ D_{2M} = -\delta_{21}\Delta F_1 - \delta_{22}\Delta F_2 \end{cases} \tag{6.18}$$

It can be seen from the above solution process that the derivation of the deformation coordination equation under the action of the residual stress field does not involve the structure of the part, so the above method is further extended to the solution of the residual stress and deformation perception of the n-time hyperstatic clamping system, as shown in Fig. 6.11 As shown, there are the following deformation coordination equations at different perception points:

$$\begin{cases} \delta_{11}\Delta F_1 + \delta_{12}\Delta F_2 + \cdots \delta_{1n}\Delta F_n + D_{1M} = 0 \\ \delta_{21}\Delta F_1 + \delta_{22}\Delta F_2 + \cdots \delta_{2n}\Delta F_n + D_{2M} = 0 \\ \cdots \\ \delta_{n1}\Delta F_1 + \delta_{n2}\Delta F_2 + \cdots \delta_{nn}\Delta F_n + D_{nM} = 0 \end{cases} \tag{6.19}$$

where ΔF_i represents the change value of the constraining force on each redundant constraint; δ_{ij} represents the displacement of the statically determinate structure along the F_i direction when $\Delta F_i = 1$ is acting alone; D_{iM} represents the displacement

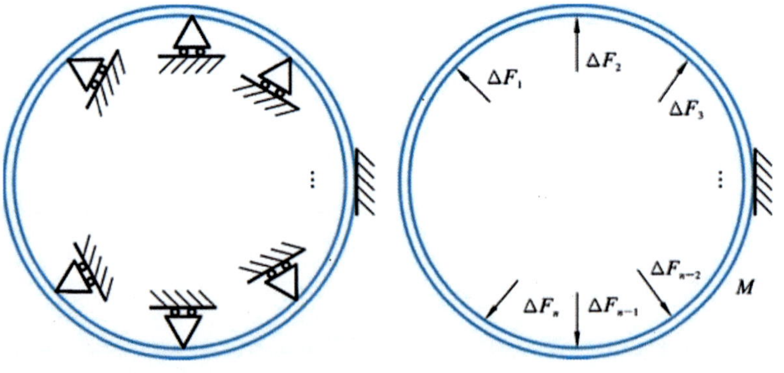

Fig. 6.11 N-time hyperstatic clamping system

of the statically determinate base structure along the F_i a direction under the sole action of residual stress.

According to the displacement reciprocity theorem, the displacement caused by the force ΔF_i at the point of application of F_j along the direction F_j is equal to the displacement caused by the force ΔF_j at the point of application of F_i along the direction F_i, that is, the coefficients in the above equations have the following relationship:

$$\delta_{ij} = \delta_{ji} \tag{6.20}$$

The application of this formula can simplify the solving of the coefficients of the equations, and the coefficients can be solved by theoretical derivation or simulation.

From the above solution process, it can be seen that the RSID perception prediction method based on clamping force monitoring can accurately obtain the RSID value of the sensing point, but the evaluation of the non-sensing point needs to be obtained by means of displacement field fitting. This means that there is a certain degree of uncertainty in the prediction effect of the residual stress and deformation of the points with a longer sensing distance. Therefore, it is necessary to plan the distribution of the sensing points reasonably, and comprehensively sense the areas with high accuracy and small deformation stiffness to ensure deformation not too bad.

Based on the above-mentioned perception model, model solution method and discussion of the perception process, the realization steps of the perception prediction of RSID are as follows.

(1) Selection of sensing points. Through simulation or theoretical analysis, the possible deformation state of the workpiece is estimated. At this time, the structure of the workpiece and the distribution of residual stress can be simplified, and the change trend can be roughly estimated, and the sensing point can be applied to the larger deformed part.
(2) Design of clamping scheme. When the redundant constraint is point contact, the residual stress and deformation of the sensing point can be accurately solved. Therefore, during the clamping design, while ensuring the stability of the clamping, try to use the point clamping form as the redundant restriction, and install the clamping force sensor on it; if the surface clamping form must be adopted, the clamping surface area should be reduced as much as possible to approximate the deformation value.
(3) Machining process perception. Clamp the workpiece and complete the processing of the workpiece according to the given working conditions. Record the clamping force at the sensing point of certain time.
(4) Deformation solution. The finite element model of the statically determinate basis of the workpiece clamping system is established, and the reaction force of the change value of the clamping force is applied at the sensing position to obtain the deformation state of the workpiece. The deformation state at this time is the residual stress change value after the workpiece is clamped and unloaded.

6.6.2 Application Cases in Thin-Walled Parts Machining

For a GH4169 material sheet, size of 160 mm × 20 mm × 2 mm, take the surface of the middle area to machine a 80 mm × 20 mm area, the cutting depth is 0.5 mm, and the area of 40 mm × 20 mm at both ends is used for clamping. Thin-walled parts undergo heat treatment before machining to release stress, and the RSID is mainly caused by the residual stress of the surface machining. The following describes how to realize the perception of residual stress and deformation of the thin-walled parts.

6.6.2.1 Perceived Location

In order to determine the sensing position, a finite element model for predicting and analyzing the MIRS induced deformation of the statically determinate base of the clamping system of thin-walled parts is established in the finite element analysis software, and the position of weak stiffness is determined by simulating the MIRS induced deformation of the part under a certain stress distribution. Determine the perception point, the process is as follows:

(1) Model establishment and setting. The three-dimensional model of the thin plate is established, and the machining surface of the wall plate is divided into 7 layers by the method of thin shell binding, and each layer is 15 μm layers, which is the effective area of the machining induced residual stress. Set the thin plate material to GH4169, one end of the thin plate is fixed and constrained, the mesh is divided by eight-node hexahedral solid elements, and the workpiece is discretized into 192,000 elements.
(2) Application of residual stress. Apply the residual stress distribution shown in Table 6.1 to each layer. This residual stress is measured by X-ray diffraction after milling the GH4169 material block with the parameters used in the experiment. If machining under variable working conditions is adopted, since the fixed tool has similar distribution characteristics of residual stress within a certain working condition range, the residual stress distribution under a certain parameter in the working condition range can be applied to the machining surface to evaluate the maximum deformation position.

(3) Calculation and post-processing of residual stress. According to the finite element calculation results, the deformation cloud image of the workpiece after milling is obtained, and the RSID state of the workpiece under static clamping is obtained through the finite element model, as shown in Fig. 6.12. It can be

Table 6.1 Residual stress distribution on machined surface

Depth (μm)	0	15	30	45	60	75	90
σ_{xx}(MPa)	−420	−584	−568	−447	−286	−133	−15
σ_{yy}(MPa)	70	−193	−234	−166	−77	−12	20

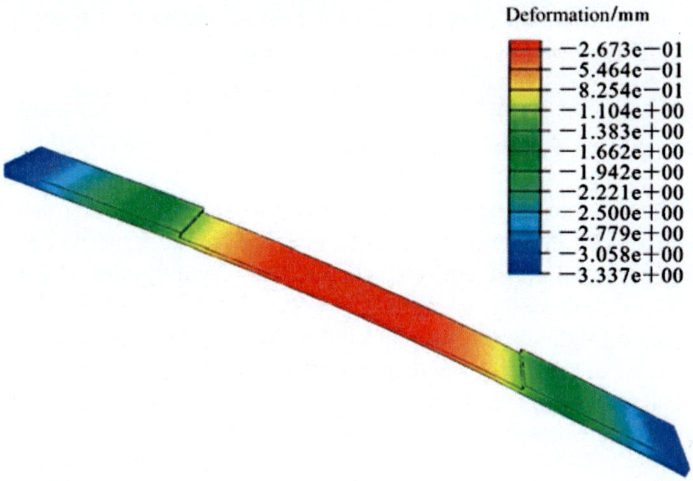

Fig. 6.12 Residual stress-induced deformation nephogram

seen that the single-sided machining of thin-walled parts shows bending deformation as a whole, and the maximum deformation position is at the boundary of the length of the workpiece. Considering the actual clamping limit, the clamping at one end and the part become a statically determinate base, and the sensing point is set at the midpoint of the width 10 mm from the boundary.

6.6.2.2 Design of Perception Fixture

The fixture design is shown in Fig. 6.13, using one end clamping and the other end point clamping. A pressure sensor is installed on the point-to-point end, and the contact surface of the sensor and the workpiece is in point contact with the arc surface. In order to ensure the accurate positioning of the point-to-point position, there are three threaded positioning holes on the corresponding positions of the upper and lower clamp parts. Together with the threaded holes on the sensor, the accuracy of the sensor installation position is ensured. In addition, in order to prevent the phenomenon of vibration loosening of the clamping during machining, the bolts are

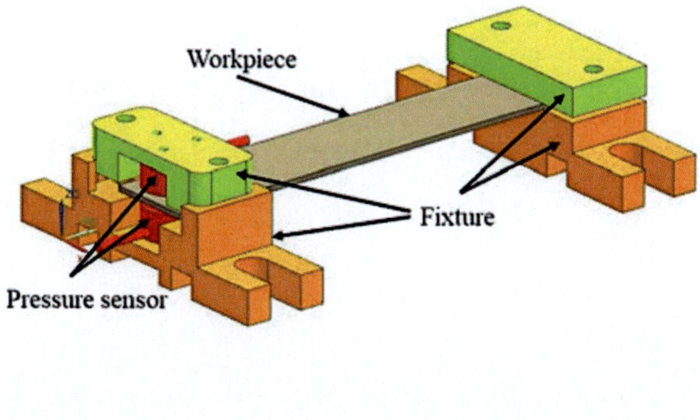

Fig. 6.13 Thin plate clamping system

fixed with self-locking nuts to ensure that the change in clamping force before and after machining is caused by the introduction of residual stress.

6.6.2.3 Machining Perception Experiment

The experiment was processed with a two-tooth flat-bottom milling cutter, with a cutting depth of 0.5 mm, a cutting width of 2 mm, and a cutting speed of 80 mm/min. Before machining, the upper sensor's value is 31.6 N and the lower sensor's value is 32.8 N. The difference of the upper and lower clamping force is caused by the linearity deviation of the sensor. After machining, the upper sensor's perception value is 31.3 N, and the lower sensor's perception value is 33.2 N. After machining, unload the sensor end fixture, and use the deformation measuring equipment to measure the deformation of the workpiece along the length center line.

6.6.2.4 Solution of Deformation

It can be seen from the sensing results that the force of the upper part of the thin-walled parts on the plate changed to −0.3 N, and the direction is downward; the force of the lower part of the part changed to 0.4 N, and the direction is upward. According to the solution formula, the RSID of the thin-walled part is equivalent to the deformation value of the downward load of 0.3 N and 0.4 N applied to the sensing position of the upper and lower surfaces of the workpiece. The finite element simulation model is established in the finite element analysis software, the downward load is applied at the sensing point to obtain the deformation cloud image of the workpiece, and the deformation value along the length center line is extracted. The comparison between the perceptual prediction result and the measured result is shown in Fig. 6.14. The measured maximum deformation is 1.07 mm, the perceived maximum deformation is 0.93 mm, and the maximum perceived prediction error is 13%.

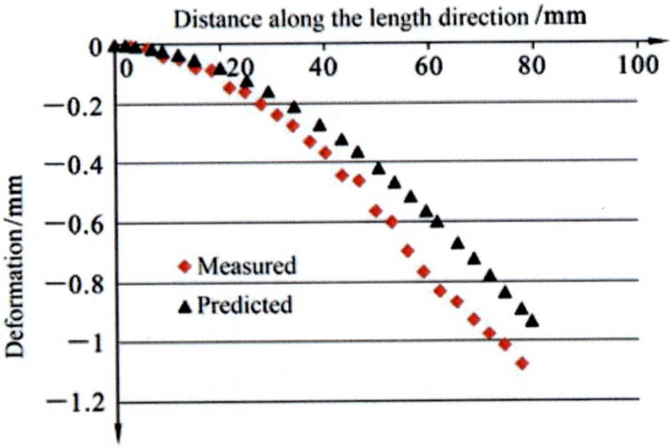

Fig. 6.14 Comparison of the perception prediction results and measurement results

6.7 Active Control Method for Residual Stresses Induced Deformation of Thin-Walled Parts

The purpose of thin-walled workpiece deformation prediction is to optimize and control the parts machining process according to the prediction results, so as to reduce the deformation and improve the machining accuracy. During the cutting process of complex thin-walled parts, the residual stresses constantly evolve. After machining, the residual stresses in the surface layer and the internal stress rebalance and lead to the deformation of the parts. At present, in the actual production, the strategy of totally loosening the fixture in-process and re-clamping the part after releasing the residual stress is widely used to control the machining deformation, which lacks the quantitative analysis and control of the RSID. In addition, the whole machining process needs to adjust the datum for many times, and at the same time, it may increase the heat treatment process which resulting in extra problems, such as complex process route, long machining cycle, high cost and poor quality stability [5].

Aiming at the deficiencies of existing control methods for RSID, according to the equivalent loads and their evolution of internal stress and MIRS for the arbitrary cutting process, the in-process deformation active control method is presented to control the RSID. The idea of this approach is to control the distribution of internal stress by actively adjusting the in-processes deformation after each machining process, in which the internal stress is balanced with MIRS. The procedure of active control method is as follows: (1) analyze the clamping force and equivalent loads of residual stresses on clamping points, (2) establish equilibrium equation of equivalent loads in-processes, (3) research the evolution of equivalent loads on each clamping points in machining and deformation process, (4) present RSID active control algorithm for thin-walled parts and achieve precision regulation to machining induced deformation.

6.7.1 *Evolution of Residual Stress in Machining Process*

6.7.1.1 Simplified Model of Parts

A workpiece is generally manufactured by a variety of machining processes, the forming processes of blank, such as forging and casting, will lead to internal stress inside the workpiece material. Then, some internal stresses will be released with the materials being removed, and the force-thermal coupling in the cutting process will generate MIRS in the surface layer of the workpiece. Therefore, the residual stresses after each machining process can be divided into two parts, the internal stress throughout the workpiece and the MIRS in the top thin layer of the component, which is denoted by σ_I and σ_m, respectively. Generally speaking, MIRS and internal stress cannot keep balance with each other. The workpiece needs to be constrained by the fixture to maintain its machined shape of the processing, which means the residual

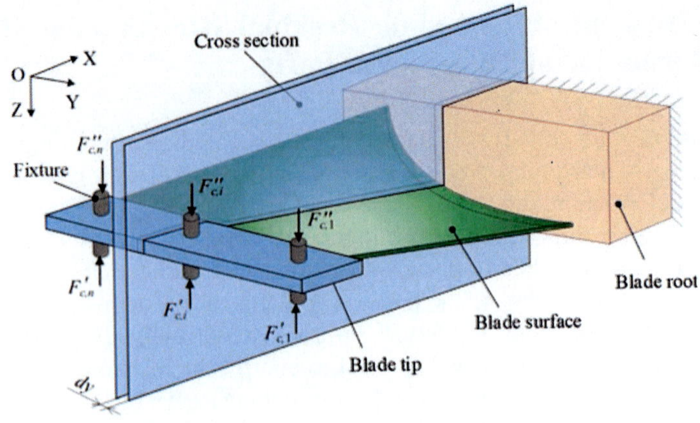

Fig. 6.15 The in-process model of blade part

Fig. 6.16 The simplified model of the in-process blade

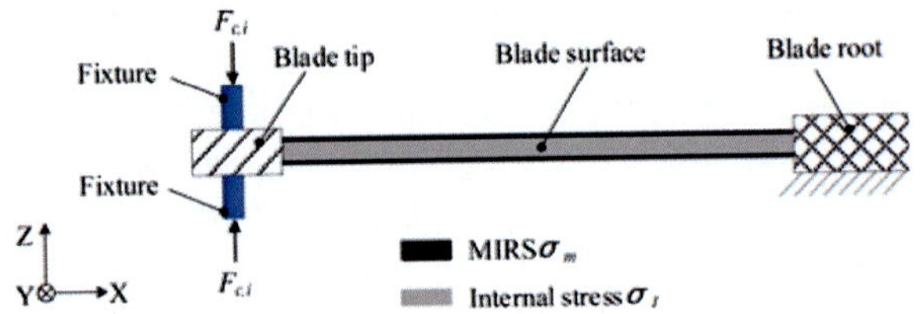

stresses are balanced with the clamping force of the fixture. Therefore, the existing clamping force reflects the imbalance of residual stresses.

Blade parts are usually processed by four-axis or five-axis CNC machining center. The machining procedures can be represented as: firstly, a specialized fixture is used to fix the blade root on the rotary shaft of the machine tool, and then the blade tip is supported by a hole on process boss to prevent the blade from being a cantilever beam, finally, the blade starts to be machined. The schematic diagram of the in-process blade-fixture system is shown in Fig. 6.15. With the help of the finite element analysis approach, the blade is divided along the O-Y direction by a plane parallel to the XZ plane, and each disk element can be regarded as a simple supported beam model, as shown in Fig. 6.16.

6.7.1.2 Equilibrium Equation of Thin-Walled Workpiece

For titanium alloy materials, the thickness of MIRS affected layer produced on the surface is usually less than 0.2 mm. Compared with other cutting methods, this affected layer generated by milling is usually smaller. Therefore, in the adjacent cutting process, the MIRS created by previous process will be completely removed in the next process, and the final MIRS on the surface layer is only influenced by the cutting parameters of the finishing process.

The machining of thin-walled parts can choose different process schemes according to the structure of the workpiece. There are mainly two types: (1) no relaxing of the fixture at all, that is, the residual stress is not released, (2) totally relaxing of the fixture in-process, that is, the residual stress is completely released, and re-clamping the part in deformation state. The following analysis is the evolution of the residual stress and equivalent loads of different machining schemes.

1. **No relaxing of the fixture at all in-process**

In the initial state, the initial internal stress inside the workpiece material is in a self-balanced state. At this time, the vector sum of the clamping forces of the part is zero. However, the original balance is broken after machining, which results in the changes of the clamping forces. Therefore, the magnitude of the clamping forces reflects the imbalance of the residual stresses. The in-process residual stress of the parts includes internal stress and MIRS. What's more, the part is also subject to the restraining force of the fixture. Under this circumstance, the clamping force on the clamping points of each slice is balanced with the equivalent loads caused by residual stress, so that the part can maintain the geometrical shape and dimensional accuracy after the machining. The equilibrium equation is achieved as:

$$F_{Ie\_i} + F_{me\_i} + F_{ce\_i} = 0(\mathrm{i} = 1, 2, \ldots, n_c) \tag{6.21}$$

where F_{Ie_i} and F_{me_i} are the equivalent loads at the ith slice caused by internal stress and MIRS, respectively. F_{ce_i} is the resultant of clamping force of the ith slice, n_c is the number of slices.

Since the residual stress of the part is not relaxed between the cutting processes, that is, the part does not deform, and the internal stress will not be redistributed. All MIRS generated in intermediate processes are removed with the machining allowance in the subsequent cutting process. Therefore, when the last process is finished, the internal stress of the part is consistent with the distribution of the initial internal stress of the blank, whereas the MIRS is only related to the parameters of the last process. After machining and the constraints of the part are removed, the internal stress and clamping force are redistributed and reach a new balance state. The internal stress and MIRS of the part are, respectively, denoted by σ'_{I_n} and σ'_{m_n}. Then, a new balance equation can be obtained as:

$$F'_{Ie\_n} + F'_{me\_n} = 0 \tag{6.22}$$

where F'_{Ie_n} and F'_{me_n} represent equivalent loads of the nth slice after deformation, that induced by internal stress and MIRS.

2. **Totally relaxing the fixture in-process**

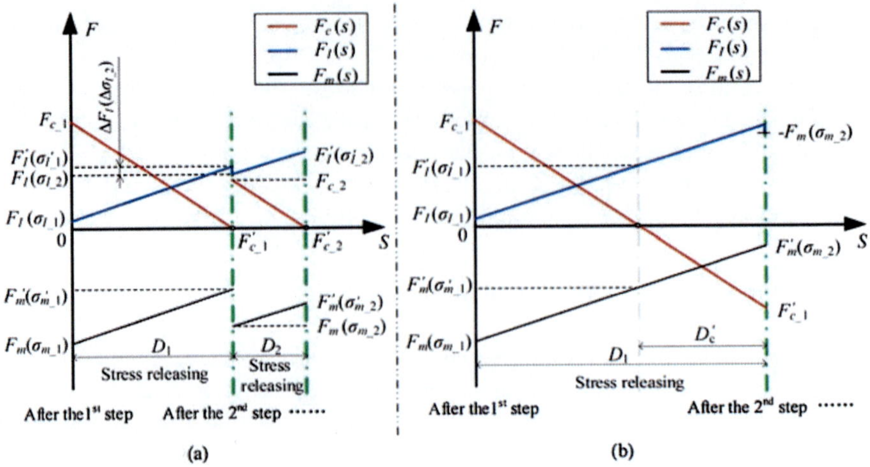

Fig. 6.17 The variation of equivalent loads: **a** totally relaxed; **b** actively controlled

The fixture is usually totally relaxed in-processes to releasing the residual stresses of the part. The clamping forces decrease gradually and become zero eventually during the fixture removal process. The internal stress and MIRS are changed and became balance with each other through deformation. The internal stress and MIRS of the deformed part are represented by σ'_{I_i} and σ'_{m_i}, respectively, and the equivalent loads relative to the clamping points are represented by $F'_{Ie,i}$ and $F'_{me,i}$. And then, the equilibrium equation can be expressed as:

$$F'_{Ie,i} + F'_{me,i} = 0(\mathrm{i} = 1, 2, \ldots, n_c) \tag{6.23}$$

where $F'_{Ie,i}$ and $F'_{me,i}$ represent equivalent loads of the ith slice after deformation, that induced by internal stress and MIRS.

The schematic diagram of Fig. 6.17a shows the variation of clamping force and equivalent loads in the machining and stress releasing process. The horizontal axis is the displacement and the vertical axis represents the equivalent load of the simplified model. Because the MIRS σ'_{m_i} of the surface layer is removed in the $i+1$th process with the removal of material, its equivalent load F'_{m_1} will disappear, too. Then, it is being replaced by the MIRS generated in the $i+1$th process, which is represented by σ_{m_i+1}. The internal stress contained in the removed material in the $i+1$th process is denoted by $\Delta\sigma'_{I\_i+1}$. At this time, the equilibrium equation of the equivalent loads on the clamping points can be expressed as:

$$F'_{Ie\_i} + \Delta F_{Ie\_i+1} + F_{me\_i+1} + F_{ce\_i+1} = 0 \tag{6.24}$$

where ΔF_{Ie_i+1} and F_{me_i+1} represent equivalent loads generated in the $i+1$th process, that induced by $\Delta\sigma'_{I\_i+1}$ and σ_{m_i+1}.

According to the elastic mechanics and material mechanics, the RSID studied in this book belongs to small-deformation and can be considered as homogeneous and linear elastic deformation. Therefore, the internal stress caused equivalent load after deformation can be expressed as:

$$F'_{Ie\_i} = F_{Ie\_i} - \frac{1}{2}\left(F_{Ie\_i} + F_{me\_i}\right) = \frac{1}{2}\left(F_{Ie\_i} - F_{me\_i}\right) \tag{6.25}$$

Based on the above analysis, the in-process deformation will lead to the redistribution of internal stress. Therefore, the distribution of the internal stress can be controlled by accurately adjusting the in-process deformation so that the equivalent loads of internal stress and MIRS are balanced with each other, which will prevent the redistribution of residual stress after the fixture is relaxed and improve the accuracy of the final parts.

6.7.2 The In-Processes Active Control Method

6.7.2.1 Deformation Superposition Principle

The RSID of the part is within the range of elastic deformation, thus, the RSIM of thin-walled parts meets the principle of deformation superposition, that is, the deformation generated by multiple loads applied to the same part is equal to the superimposed result of the deformation caused by each load separately. After finishing each process, loads of the part in the fixed state include the clamping force, the MIRS of the surface layer, and the internal stress. Their combined action makes the workpiece maintain the machined shape, that is, the deformation of the part in the clamping state is zero. The schematic diagram of deformation superposition is shown in Fig. 6.18.

Therefore, the totally deformation of the part is equal to the algebraic sum of the deformation that caused by the equivalent load of MIRS, internal stress and clamping force, which lead to the equation:

$$D_{is} + D_{ms} + D_c = 0 \tag{6.26}$$

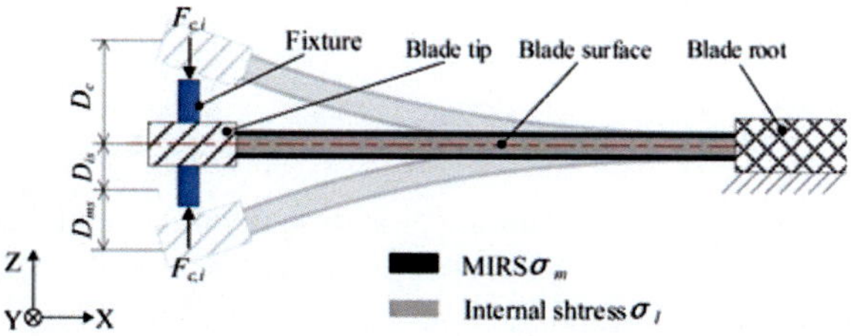

Fig. 6.18 The schematic of the deformation superposition principle

where D_{is} denotes the deformation of clamping point caused by the internal stress σ_I, D_{ms} by σ_m, and D_c by F_c.

6.7.2.2 Active Control Method

1. **No relaxing the fixture at all in-process**

The internal stress of the final component is equal to the initial stress of the blank if the component is fixed during the whole machining process. Under this circumstance, the deformation of the final component is determined by the internal stress of component σ_{I_n} and MIRS of the last machining operation σ_{m_n}. According to the deformation superposition principle, the final deformation of the part can be expressed as:

$$D_n = D_{Is\_n} + D_{ms\_n} \tag{6.27}$$

where D_n denotes the deformation of the final component, D_{Is_n} and D_{ms_n} are the deformation caused by σ_{I_n} and σ_{m_n}, respectively.

2. **Totally relaxing the fixture in-process**

However, the deformation will be different if the fixture is totally relaxed to releasing the residual stresses after each machining step. The internal stress and MIRS after the first step are represented with σ_{I_1} and σ_{m_1}, respectively. The deformation after the first step can be expressed as:

$$D_1 = D_{Is\_1} + D_{ms\_1} \tag{6.28}$$

where D_1 is the deformation after the first step, D_{Is_1} and D_{ms_1} are the deformation caused by σ_{I_1} and σ_{m_1}, respectively.

After the residual stress is completely released, the MIRS and the internal stress of the part will be redistributed. Internal stress denoted as σ'_{I_1} is balanced with the MIRS denoted as σ'_{m_1} after the part deforms. Then, the RSID of the part after completely releasing the restraint of the fixture:

$$D'_{Is\_1} = -\frac{1}{2}\left(D_{Is\_1} + D_{ms\_1}\right) \tag{6.29}$$

where D'_{Is_1} denotes the deformation caused by σ'_{I_1}.

The second machining operation is implemented after the part re-clamped at the new state, in which the residual stresses are rebalanced completely. When the second process is finished, the internal stress and MIRS are represented by σ_{I_2} and σ_{m_2}. The deformation can be described as:

$$D_2 = D_{Is\_2} + D_{ms\_2} \tag{6.30}$$

where D_2 is the deformation after the second process, D_{ms_2} is the deformation caused by σ_{m_2} and D_{Is_2} by σ_{I_2}.

It can be seen from the analysis in the previous section that the variation of internal stress caused by the material removal in the second process is recorded as $\Delta\sigma_{I\_2}$. Therefore, after the end of the second process, the deformation caused by the internal stress alone can be expressed as:

$$D_{Is\_2} = \left(D'_{Is\_1} + \Delta D_{Is\_2}\right) \tag{6.31}$$

where D'_{Is_1} is the deformation caused by σ'_{I_1} and ΔD_{Is_2} by $\Delta\sigma_{I\_2}$.

Similarly, the deformation of the part after any process i can be calculated, as shown in Eq. (6.32). The internal stress and MIRS are represented by σ_{I_i} and σ_{m_i}, respectively.

$$D_i = D_{Is\_i} + D_{ms\_i}(0 < i < n) \tag{6.32}$$

where D_{Is_i} is the deformation caused by σ_{I_i} alone when the ith process finished, and D_{ms_i} by σ_{m_i}. D_i is the total deformation after the ith process. n is the number of the process when forming.

Then the deformation caused by the internal stress after the ith process can be calculated through Eq. (6.31).

$$D_{Is\_i} = D'_{Is\_i-1} + \Delta D_{Is\_i} \tag{6.33}$$

where ΔD_{Is_i} is the deformation caused by $\Delta\sigma_{I\_i}$ which is produced due to the material removal, D'_{Is_i-1} denotes the deformation caused by σ'_{I_i-1}.

Substitute Eq. (6.33) into Eq. (6.32), the deformation after any process is expressed as:

$$D_i = D'_{Is\_i-1} + \Delta D_{Is\_i} + \Delta D_{ms\_i} \tag{6.34}$$

According to Eq. (6.29), D'_{Is_i-1} can be denoted as:

$$D'_{Is\_i-1} = -\frac{1}{2}\left(D_{Is\_i-1} + D_{ms_i-1}\right) \tag{6.35}$$

where D_{Is_i-1} is the deformation caused by internal stress of the part after i-1th process and D_{ms_i-1} by MIRS of i-1th process.

It can be seen from Eq. (6.32) to Eq. (6.35) that the deformation of the final component will be affected by the residual stress and MIRS if the fixture is relaxed after each process. Therefore, the completely release of the residual stresses in-process will cause the internal stress of the part to change continuously. However, when the MIRS generated by different processes are significantly diverse, the internal stress of the part will change greatly. At the same time, the balance of the internal stress and MIRS cannot be guaranteed, and the machining deformation of the final component could not be eliminated.

3. **Actively controlling the deformation in-process**

The realization principle for the in-process active control method of RSID is shown in Fig. 6.19, where D_i represents the deformation when the fixture is relaxed after the end of the ith process, D'_{c_i} represents the deformation produced by adjusting the clamping force, which is called active control deformation. After the end of the ith process, the clamping force F'_{c_i} is introduced into the workpiece-fixture system, and the deformation can be expressed as:

$$D''_{Is\_i} = D'_{Is\_i} + D'_{c\_i} = -\frac{1}{2}\left(D_{Is\_i} + D_{ms\_i}\right) + D'_{c\_i} \tag{6.36}$$

where D''_{Is_i} is the deformation after active control, D'_{c_i} is the deformation caused by the clamping force F'_{c_i}.

In this way, the residual stresses of the component are changed and the variation of the equivalent loads is schematically shown in Fig. 6.17b. Then, after the $i + 1$th process, the deformation can be described as:

$$D_{i+1} = D''_{Is\_i} + \Delta D_{Is\_i+1} + D_{ms\_i+1} \tag{6.37}$$

If the equivalent force F_{Ie_i+1} generated by the internal stress and the equivalent force F_{me_i+1} produced by the MIRS are equal and opposite to each other, the part will not deform at all, that is, $D_{i+1} = 0$. At this time, according to Eq. (6.36) and Eq. (6.37), the in-process active control deformation can be calculated as:

$$D'_{c\_i} = -\left(D'_{Is\_i} + \Delta D_{Is\_i+1} + D_{ms\_i+1}\right) \tag{6.38}$$

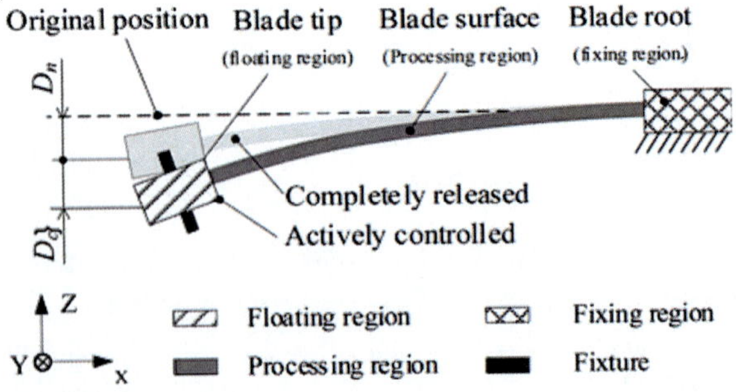

Fig. 6.19 Schematic of in-processes active control method

In the 2D illustrative example shown in Fig. 6.19, the workpiece is divided into the fixing region, the processing region, and the floating region. The fixing region is completely fixed through the machining process. The processing region is defined as the area where excess material is being removed. The floating area is fixed by the fixture when machining to improve the clamping rigidity of the blade and can be loosened or adjusted during the cutting process to release the strain energy of the part. The in-processes active control method can be realized by adjusting the fixture in the floating region between processes, and the actively controlled deformation can be calculated.

6.7.3 *Application of Active Control Method for RSID in Blade Machining*

The aforementioned method is adopted to actively control the RSID during the processing. Design different machining schemes to compare the deformations and verify the accuracy and precision of in-process active control method. These process schemes are no relaxing the residual stress at all, totally relaxing the residual stress in-process and actively controlling the deformation in-process. The three group experiments include the rough milling and the semi-finish milling and have the same cutting parameters and working conditions. To realize the presented in-processes active control method, a prototype of active control fixture for the blade part is developed. The CAD model and the real fixture are shown in Fig. 6.20.

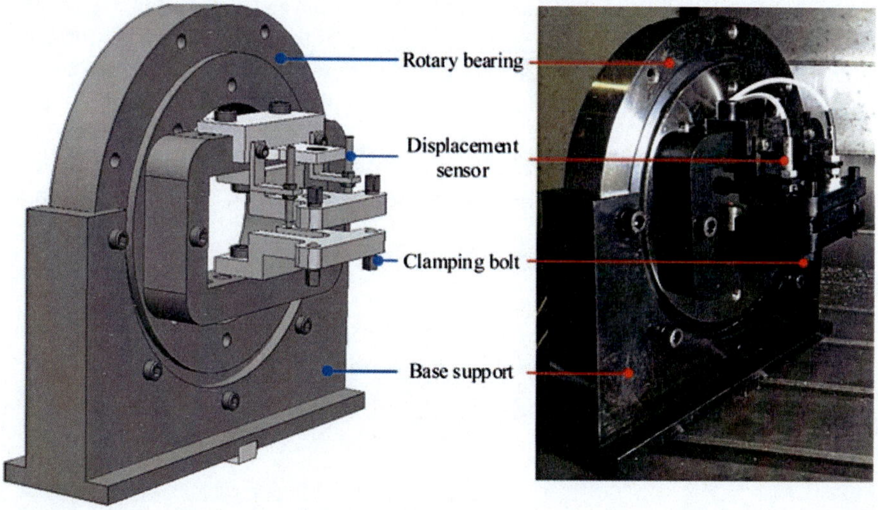

Fig. 6.20 The prototype of active control fixture

In order to quantitatively analyze the experimental results, four measuring cross-sections, denoted by A, B, C and D, are placed at X = 35 mm, 65 mm, 95 mm and 125 mm, respectively, on the workpiece coordinate system O_XYZ. There are 40 measuring points on each cross-section and the measurement trajectory is as shown in Fig. 6.21. The processing deformation of the blade is measured by the on-machine measurement instrument RENISHAW RMP600 and the radius of the measuring head is 3 mm.

The regulating methods of these three experiments are as follows:

(1) **The first group (i.e., no relaxing the fixture at all).** The residual stresses are not released after the rough milling and the semi-finish milling is performed immediately after the rough milling without removing/changing the fixture. The active control fixture is relaxed after semi-finish milling. The deformations of the measuring cross-sections and the deformations of the observation points (denoted as D_2^1) are recorded, respectively.

(2) **The second group (i.e., totally relaxing the fixture in-process).** The active control fixture is relaxed completely to rebalance the residual stresses after the rough milling and the deformations of observation points are recorded (denoted as D_1^2). The workpiece is then re-clamped at the new state after the residual stresses are redistributed. After that, the semi-finish milling is performed. Finally, the active control fixture is relaxed, and the deformations of the measuring cross-sections and the deformations of the observation points (denoted as D_2^2) are recorded, respectively.

(3) **The third group (i.e., actively controlling the deformation in-process).** Firstly, the finite element prediction method of MIRS induced deformation is used to predict the MIRS induced deformation after the rough milling and semi-finish milling of the blade, which are recorded as D_{ms_1} and D_{ms_2}, respectively. After the rough milling, relax the bolts of the active control fixture, and record the deformation of the observation point, which is presented as D_1^3. Use the proposed active control algorithm to calculate the deformation of each observation point, which is recorded as D'_{c_1}. Before the semi-finishing, re-clamp the workpiece according to the in-process active control deformation method and then perform the semi-finishing. After that, loosen the clamping

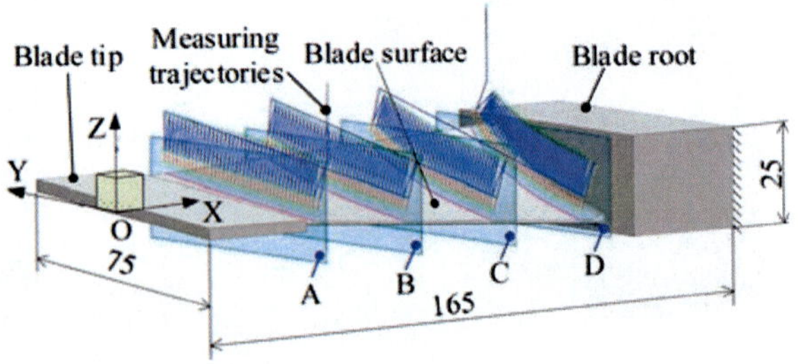

Fig. 6.21 The blade model and measuring trajectories

bolts of the fixture and measure the deformation of each measuring cross-section, and record the deformation on the observation points OA and OB, which are recorded as D_2^3.

The displacements of observation points in the three group experiments are shown in Table 6.2.

The deformation data and fitting results of each measuring cross-section of the three blades obtained from the experiments are shown in Fig. 6.22. For the first group, the final deformation of the part is mainly affected by the MIRS generated by the semi-finishing process. As shown in Fig. 6.22a, the blade has severe bending and torsion deformation after the semi-finishing. For the second group, as shown in Fig. 6.22b, the difference of the MIRS magnitude generated by the rough milling and semi-finish milling is relatively small. Therefore, the equivalent force of the internal stress is almost equal to the equivalent force of the MIRS, then the deformation of the part decreases, too. For the third group, as shown in Fig. 6.22c, after the rough milling, the clamping force is introduced at the tip of the blade to control the deformation and change the distribution of the internal stress of the part, so that when the semi-finishing is completed, the equivalent force of the internal stress and the MIRS are theoretically equal. Thus, the redistribution of internal stress and MIRS, which will cause deformation can be avoided after the semi-finishing is completed, and the machining deformation of the final component is significantly reduced. For the measuring cross-section A of the three experiments, compared with no relaxing the fixture at all, totally relaxing the fixture and in-process active control method can reduce the deformation by 57.1% and 81.7%, respectively.

Table 6.2 The displacements of observation points

Observation points	The first group	The second group		The third group		
	$D_2^1(um)$	$D_1^2(um)$	$D_2^2(um)$	$D_1^3(um)$	$D'_{c_1}(um)$	$D_2^3(um)$
OA	155	186	84	178	86	34
OB	2	−39	9	−32	22.5	−3

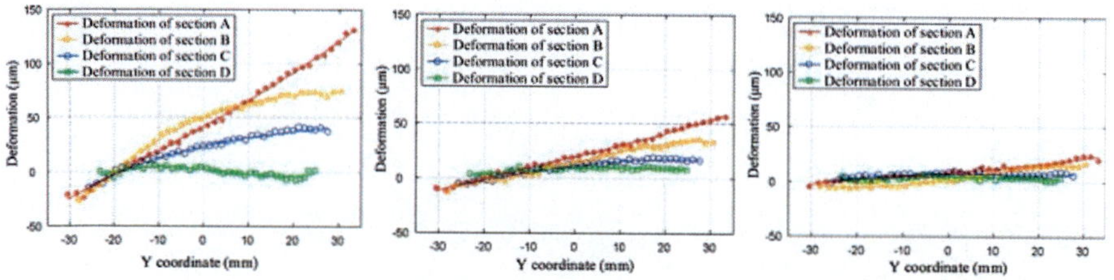

Fig. 6.22 Deformation results of **a** the first group **b** the second group **c** the third group

References

1. WANG J T. Research on prediction and optomization method of deformation induced by residual stresses in milling of thin-walled parts [D]. Xi'an; Northwestern Polytechnical University, 2019.
2. WANG J T, ZHANG D H, WU B H, et al. Prediction of distortion induced by machining residual stresses in thin-walled components [J]. The International Journal of Advanced Manufacturing Technology, 2018, 95(9): 4153–4162.
3. YANG Y F. Research on machining deformation prediction and control technology of monolithic structure based on internal stress field [D]. Nanjing; Nanjing University of Aeronautics and Astronautics, 2010.
4. ZHANG P X. Theoretical structural mechanics of energy [M]. Shanghai: Shanghai Scientific & Technical Publishers, 2010.
5. ZHANG Z X. The prediction and control methods for the deformation of complicated thin-walled parts that induced by residual stress [D]. Xi'an; Northwestern Polytechnical University, 2021.